AF522277

GREEN PERSPECTIVES IN FOOD PROCESSING

a division of
NIPA GENX ELECTRONIC RESOURCES & SOLUTIONS P. LTD.
New Delhi 110 034

About the Editors

Dr. Anupama Singh has over 27 years of academic experience in the agri-food processing sectors. Her research interest includes Bio waste utilization, Sustainable food processing novel technologies, product development and value addition.

Dr. Singh has received various accolades, recognitions and fellowships and awards at the national and international levels, including the prestigious Norman Borlaug Fellowship by USDA/ICAR and the National Fellow Award by ICAR, India.

She has executed multiple R&D and Consultancy projects. She has guided 4 doctoral research, 28 M. Tech. Thesis and many research projects at the graduate level.

She has over 250 publications to her credit including research papers/articles in various peer-reviewed international and national journals, book chapters, technical bulletins, status reports, articles etc. She has presented more than 100 research papers in various National &International conferences.

After a sterling career, spanning over 25 year, at GB Pant University of Agriculture & Technology, Pantnagar, Dr. Anupama Singh is currently Professor and Head at Department of Food Engineering, National Institute of Food Technology Entrepreneurship and Management (NIFTEM), India.

Dr. Komal Chauhan had the rare opportunity of being nurtured in varied regions and cultures of India bringing out the best in her. She did her doctorate in Food Science and Nutrition from Banasthali University Rajasthan. A meritorious student throughout her academic career she has been a scholarship holder and Gold Medalist at Masters Level. She embarked on her teaching career from SD College Ambala Cantt and later to Banasthali University. At present she is working as Associate Professor in Nutrition in the Dept. of Food Science and Technology at National Institute of Food Technology Entrepreneurship and Management, Kundli, Sonipat, Haryana, India. She has a teaching and research experience of more than two decades. She has published and presented several papers in National and International Journals and conferences. She has been working in the area of Nutraceuticals and

Functional Foods and Nutritional Biochemistry. She has been working in the areas of malnutrition, and non-communicable diseases namely diabetes, cardiovascular diseases and obesity. She has worked on various projects sponsored by UNICEF, UGC, DST/SERB/MoFPI and Ministry of Tribal Affairs, MP Govt, WCD Govt. of Kerala.

Dr Rajni Chopra currently working as Associate professor in the Department of Food Science and Technology, NIFTEM, Haryana, India. She has more than 13 year teaching and research experience. She completed her PhD degree in Biochemistry in the year 2008 from Central Food Technological Research Institute (CFTRI), Mysore, Karnataka in 2008. She Received UGC JRF for pursuing her Doctoral Research and also received ICAR JRF for pursuing Master degree. She has worked as Assistant Professor IHE, University of Delhi for 12 years. She has published more than 20 research papers in international and national journals, and published 4 book chapters and one process patented. She has guided more than 26

M.Sc. dissertations and guiding 4 PhD research students presently. She has delivered talks in the international and national conferences and presented various papers at national and international conferences. Her area of interest is Lipid Science and Technology, modification of edible oils through various techniques and oil based nutraceuticals and by-product utilization.

Dr Rakhi Singh is Assistant Professor in the Department of Food Science and Technology, National Institute of Food Technology Entrepreneurship and Management (NIFTEM), Haryana, India. She has more than 15 year teaching and research experience. She has worked as Assistant Professor in Centre of Food Science & Technology, Banaras Hindu University, and Sam Higginbottom University of Agriculture, Technology, and Sciences, Allahabad, India. She has guided more than 23 MSc and M. Tech dissertations, 3 PhD scholars and presently guiding 3 PhD research scholars. She has published more than 32 research papers and book chapters in the journals/ books of national and international repute. She has delivered talks and presented various papers at national and international conferences. Her present areas of interest include health foods, nutraceuticals and traditional food products.

Dr. Anurag Singh is working as Assistant professor in the department of Food Science and Technology at National Institute of Food Technology Entrepreneurship and Management (NIFTEM), Sonipat, India. Dr. Singh has teaching experience of more than 15 years and is actively involved in research in the field of food technology. His main areas of research are fruits and vegetable processing, bakery technology, and new product development. He has published more than 25 research papers and book chapters in the journals/ books of national and international repute. He has also handled different sponsored projects funded by various funding agencies like DST, MoFPI, government agencies like KVIC, EDMC as well as reputed industries such as Jubilant Agri and Consumer Pvt Ltd., Pepsico India Holdings Pvt. Ltd. Dr. Singh has completed his B.E. in Food Technology and Engineering from Faculty of Engineering and Technology, R.B.S. College, Agra (India); M.Tech. (Food Technology) from Harcourt Butler Technological Institute, Kanpur (India) and PhD in Food Technology from Mahatma Gandhi Chitrakoot Gramodaya Vishwavidyalay, Satna (India). Dr. Singh is the life member of The Institution of Engineers (India); Indian Science Congress; Association of Food Scientists & Technologists (India) and Indian Society of Genetics, Bio-Technology Research & Development (ISGBRD).

Er. Pramod K Prabhakar is an Assistant Professor in the Department of Food Science and Technology, NIFTEM Kundli, Haryana. He graduated in Food Process Engineering specialization from Indian Institute Technology, Kharagpur. He has received GATE and UGC Fellowships for M. Tech and PhD studies and qualified ASRB-NET. Er. Prabhakar has also worked as an Assistant Professor in the Department of Food Engineering, NIFTEM in 2013 and thereafter worked as a Research Fellow on DBT sponsored project at IIT Kharagpur. He has been teaching various subjects at UG and PG levels such as Novel Food Processing Technologies, Fish Process Engineering, Transport Phenomena and Grain Storage Engineering. His major areas of research are Process Engineering, Transport Properties, and Food Functionalization. He has 24 publications in National and International journals, 12 book chapters, 01 book and filed one patent. Er. Prabhakar has several attended national and international conferences, symposia and seminars in India and abroad. He is a life member of AFST (India), Mysore, ISAE Delhi, Institution of Engineers, Kolkata and Indian Society for Spices, Kozhikode, Kerala

Dr. Shruti Shukla is currently working as DBT-Ramalingaswamy Faculty Fellow in the Department of Food Science and Technology at NIFTEM. Dr. Shruti Shukla is having more than 10 years abroad experience as Assistant Professor in the field of Food Science and Biomaterials at South Korean Universities. Her research interests are to explore rapid immuno-detection techniques, Nano-sensors for food pathogens and toxins, Bio-receptor developments, Oriental Fermentation, Functional foods, Food Safety. She has published more than 100 articles in the worlds' most prestigious SCI indexed journals. She is also serving editorial role in various prestigious international journals. Being research expertise in Nano Food Biotechnological aspects offering research ideas in multidisciplinary aspects of nano-food sciences for developing nano-sensors, nano-markers and other nanotechnological applications for food science advancements. Her continuous research interests are to explore new techniques with nano-micro structures for fabricating sensing and food toxicant adsorbing devices with the use of biomass. She has completed few International research projects in the field of fermented food and their safety aspects and for developing multiplexed portable sensor devices.

GREEN PERSPECTIVES IN FOOD PROCESSING

Anupama Singh
Komal Chauhan
Rajni Chopra
Rakhi Singh
Anurag Singh
Pramod Prabhakar
Shruti Shukla

National Institute of Food Technology Entrepreneurship and Management
NIFTEM, Kundli, Sonipat, Haryana, India

a division of
NIPA GENX ELECTRONIC RESOURCES & SOLUTIONS P. LTD.
New Delhi-110 034

a division of
NIPA GENX ELECTRONIC
RESOURCES & SOLUTIONS P. LTD.
101,103, Vikas Surya Plaza, CU Block
L.S.C.Market, Pitam Pura, New Delhi-110 034
Ph : +91 11 27341616, 27341717, 27341718
E-mail:newindiapublishingagency@gmail.com
www: www.nipabooks.com

For customer assistance, please contact
Phone: + 91-11-27 34 17 17 Fax: + 91-11- 27 34 16 16
E-Mail: feedbacks@nipabooks.com

ISBN: 978-81-94849-51-3

Composed and Designed by NIPA.

National Institute of Food Technology Entrepreneurship and Management (Deemed to be University)

(UNDER MINISTRY OF FOOD PROCESSING INDUSTRIES, GOVERNMENT OF INDIA)

Dr. Chindi Vasudevappa
Vice Chancellor

Foreword

Going green is important to minimize the effects of global warming and green technology lies in reducing the risks posed by environmental damage and conserving natural resources. It will also ensure that clean and renewable sources of energy are used to prevent the complete exhaustion of the other non-renewable sources. Green technology contributes to the various factors such as recycling the waste, purification of water and air, energy conservation and ecosystem rejuvenation. Green Food Processing leads in enhancement shelf life and the nutritional quality of food products, while at the same time reducing energy use and unit operations for processing, eliminating wastes and by-products, reducing water use in harvesting, washing and processing, and using naturally derived ingredients.

Recently, there has been a paradigm shift in the food processing technology and focus is on adopting green food processing methods in food processing industries. Due to the rapid development of food and peripheral technology, the use of new physical processing or auxiliary processing methods can maintain food inherent nutrients, texture, colour, and freshness and reduce environmental pollution and energy consumption in food processing.

I am extremely gratified that the organizers of FDP have made efforts to bring the ideas of the participants together in the form of book in which chapters are justifying the theme of the FDP on Green Prospective in Food Processing. 1 congratulate all members who were involved in giving the shape to this Idea of compiling best assignment in the form of book chapters and to collate it in the form of edited book.

Chindi Vasudevappa

Plot No. 97, Sector-56, HSIIDC Industrial Estate, Kundli - 131028, Distt. Sonepat (Haryana)
Phone: 0130-2281001-4, Fax : 0130-2219749, Email: vc@niftem.ac.in Website: www.niftem.ac.in

Preface

Dynamic changes in the environment greatly affect the Agriculture food value chain and Food operations play a vital role in the organization of the food sector as it is the key tool for transition to a green economy.

Presently, in the current pandemic situation coupled with climate changes and resource limitations, the food consumption and lifestyle of consumers are undergoing major changes and these affect the food supply systems, especially when one considers their impact on the ecological systems. Reducing this impact on the environment and maintaining a healthy standard, on a global scale, is a major challenge. Most Food industries in the current age face this challenge and the aforementioned troubles require a solution that not only maintains the standard of production but also ensures the protection of the environment.

This challenge is met by employing green strategies and the greening of the food processing sector is crucial in meeting these challenges. With the aid of these green approaches, concerns like waste generation, climate change, greenhouse emissions, energy consumptions, toxic chemicals, and carbon and water footprints are properly tackled. These Green strategies can have a direct and a significant impact on the environment and it could provide quite an effective approach for achieving sustainable systems, which would ensure growth of society along with protection to the environment. These Greener and Cleaner alternatives must be opted by the industries.

The Objective of this book is to explain and discuss the various green processing strategies, so the Book has been divided into distinct sections which cover 5 sub themes i.e Green technologies in Food Production, Post Harvest Management and Value Addition for Sustainable Value Chains, Green and eco friendly techniques in food packaging, Waste management and valorization.

The First section aims at obtaining insight in various green technologies like non thermal processing technologies. In the Second section, technologies related to waste valorization are discussed, and in Third Section, post harvest management and value addition are described in detail. The Last Section deals with the concerns of various stakeholders in this food value chain and also few topics related to eco friendly techniques.

The editors hope that the readers of this book will get a broad overview on all the relevant topics and technologies, and will also find valuable information on Green perspectives in the Food Processing Sector. This book is a welcome resource for Food Engineers/Food Scientists/Food Technologists and academia.

Key Features of the book are –

- Serves as a resource for anyone dealing with Food

- Addresses the issues and solutions to environmental problems

- Coverage of variety of topics related to Greening of processing sector

Editors

Contents

***Corresponding Author**

***Corresponding Author**

***Corresponding Author**

*Corresponding Author

***Corresponding Author**

***Corresponding Author**

List of Contributors

Chapter 1. Green Technology in Bakery Industry

S.P. Divekar* *Associate Professor of Processing and Food Engineering, Dr. Panjabrao Deshmukh Krishi Vidyapeeth, Akola-444104, Maharashtra*

V.T. Atkari *Assistant Professor, Department of Agricultural Engineering, Don Bosco College of Agriculture Solcorna, Quepem, Goa-403705*

Chapter 2. Ozone Treatment: The Green Technology in Food Industry

Krantidip R. Pawar *Assistant Professor, Department of PFE, Dr. D.Y. Patil, CAET, Kolhapur, Maharashtra*

Vinod T. Atkari* *Assistant Professor, Don Bosco College of Agriculture Solcorna, Quepem, Goa-403705*

Dilip A. Pawar *CAIE, Bhopal, Madhya Pradesh*

Pravin D. Ukey *HoD, Department PFE, Dr. D. Y. Patil CAET Talsande. Kolhapur, Maharashtra*

Chapter 3. Application of Green and Eco-Friendly Techniques in Food Packaging

Amit Pratap Singh *P.G. Department of Food Technology, R.B.S. Engineering Technical Campus, Bichpuri, Agra, Uttar Pradesh*

Anurag Singh* *Department of Food Science & Technology, National Institute of Food Technology Entrepreneurship & Management (NIFTEM), Kundli, Sonipat-131028, Haryana*

Ashish Khare *P.G. Department of Food Technology, R.B.S. Engineering Technical Campus, Bichpuri, Agra, Uttar Pradesh*

Chapter 4. Entrepreneurial Opportunities Through Value Addition of Jackfruit (*Artocarpus Heterophyllus*)

Sapna Arora* *Department of Food Business Management & Entrepreneurship Development, National Institute of Food Technology Entrepreneurship & Management (NIFTEM) Kundli, Sonipat-131028, Haryana*

***Corresponding Author**

Anjali Shankhwar *Department of Food Business Management & Entrepreneurship Development, National Institute of Food Technology Entrepreneurship & Management (NIFTEM), Kundli, Sonipat-131028, Haryana*

Chapter 5. Ultrasound Assisted Drying and Its Impact on Bioactive Compounds of Fruits and Vegetables

Swati Mitharwal *Department of Food Science and Technology, National Institute of Food Technology Entrepreneurship & Management (NIFTEM), Sonipat-131028, Haryana*

Sachin Kumar *Department of Food Engineering, National Instttute of Food Technology Entrepreneurship & Management (NIFTEM), Kundli, Sonipat-131028, Haryana*

Neerja Usha Kujur *Department of Basic and Applied Sciences, National Institute of Food Technology Entrepreneurship & Management (NIFTEM), Kundli, Sonipat-131028, Haryana*

Komal Chauhan* *Department of Food Science and Technology, National Institute of Food Technology Entrepreneurship & Management (NIFTEM), Kundli, Sonipat-131028, Haryana*

Prabhat Kumar Nema *Department of Food Engineering, National Institute of Food Technology Entrepreneurship & Management (NIFTEM), Kundli, Sonipat-131028, Haryana*

Chapter 6. Green Approach in Food Processing and Production Through Nanotechnology

Aryasree Sukumar *PhD Research Scholar, Department of Food Process Engineering, SRM Institute of Science and Technology, Kattankulathur, Chengalpattu-603203, Tamil Nadu*

K.A. Athmaselvi* *Associate Professor, Indian Institute of Food Processing Technology, Pudukkottai Road, Thanjavur-613005, Tamil Nadu*

Chapter 7. Valorization of By Products in Agro Commodity Processing Sectors

Jayati Pal Chattopadhyay* *Techno Main Salt Lake, Kolkata, West Bengal*

Aditi Roy Chowdhury *Techno Main Salt Lake, Kolkata, West Bengal*

Chapter 8. Various Microencapsulation Process Technologies Used in Food Industries: An Overview

Sumit Sudhir Pathak* *Department of Food Process Engineering, National Institute of Technology, Rourkela, Odisha*

Mahipal Singh Tomar *Department of Food Process Engineering, National Institute of Technology, Rourkela, Odisha*

***Corresponding Author**

Chapter 9. Malting and Its Effect on Nutritive Value of Food

Rakhi Singh* *Department of Food Science & Technology, Institute of Food Technology Entrepreneurship & Management, Kundli, Sonipat-131028, Haryana*

Vandana Kaushal *Department of Food Science & Technology, Institute of Food Technology Entrepreneurship & Management (NIFTEM), Kundli, Sonipat-131028, Haryana*

S. Thangalakshami *Department of Food Engineering, National National Institute of Food Technology Entrepreneurship & Management (NIFTEM), Kundli, Sonipat-131028, Haryana*

Anurag Singh *Department of Food Science & Technology, National Institute of Food Technology Entrepreneurship & Management, Kundli, Sonipat-131028, Haryana*

Chapter 10. An Insight of Cold Plasma Processing and its Application in Food Industry

Drishti Kadian *PhD. Research Scholar, Dairy Technology Division, ICAR- National Dairy Research Institute, Karnal-132001, Haryana*

Chander Mohan* *PhD. Research Scholar, Dairy Technology Division, ICAR- National Dairy Research Institute Karnal-132001, Haryana*

Chapter 11. Cooked Food Waste Management: An Eco-friendly Approach

K. Sai Shiva *Department of Food Science and Technology, National Institute of Food Technology Entrepreneurship & Management (NIFTEM), Kundli, Sonipat-131028, Haryana*

Anurag Singh* *Department of Food Science and Technology, National Institute of Food Technology Entrepreneurship & Management (NIFTEM), Kundli, Sonipat-131028, Haryana*

Rakhi Singh *Department of Food Science and Technology, National Institute of Food Technology Entrepreneurship & Management (NIFTEM), Kundli, Sonipat-131028, Haryana*

S. Thangalakshmi *Department of Food Engineering, National Institute of Food Technology Entrepreneurship & Management, Kundli, Sonipat-131028, Haryana*

Chapter 12. Green Packaging and Utilization of Fruit Fibres: A Review

Srishti Khurana *Graduate Scholar, Lady Irwin College, University of Delhi, New Delhi-110001, India*

Yashmita Grover *Graduate Scholar, Lady Irwin College, University of Delhi, New Delhi-110001, India*

***Corresponding Author**

Aparna Agarwal* *Department of Food and Nutrition, Lady Irwin College, University of Delhi, New Delhi-110001, India*

Abhishek Dutt Tripathi *Department of Dairy Science and Food Technology, Institute of Agricultural Sciences, Banaras Hindu University, Varanasi. -221005, Uttar Pradesh, India*

Ms. Anjana Kumari *Department of Food and Nutrition, Lady Irwin College, University of Delhi, New Delhi-110001, India*

Chapter 13. Processing of Banana Flower to Value Added Products: A Review

Pooja Jha *Department of Food Science and Technology, National Institute of Food Technology Entrepreneurship and Management (NIFTEM), Kundli, Sonipat-131028, Haryana, India*

Murlidhar Meghwal* *Department of Food Science and Technology, National Institute of Food Technology Entrepreneurship & Management, Kundli, Sonipat-131028, Haryana, India*

Pramod K. Prabhakar *Department of Food Science and Technology, National Institute of Food Technology Entrepreneurship & Management, Kundli, Sonipat-131028, Haryana India*

Chapter 14. Carbon Foot Printing and Green Technology: An Environment Friendly Approach in Dairy Sector

Partha Pratim Debnath* *Assistant Professor, Faculty of Dairy Technology, West Bengal University of Animal and Fishery Sciences (WBUAFS), Mohanpur-741252, Nadia, West Bengal, India*

Syed Mansha Rafiq *Assistant Professor, Department of Food Science and Technology, National Institute of Food Technology Entrepreneurship and Management (NIFTEM), Sonipat, Haryana India*

Nandini Dutta *Assistant Professor, Department of Food Technology, Faculty of Engineering and Technology, Jain University (Deemed-to-be-university), Bengaluru, Karnataka, India*

*Corresponding Author

1

Green Technology in Bakery Industry

S.P. Divekar **and** ***V.T. Atkari***

ABSTRACT

The global market for baked products has reached up to \$495 billion in the year 2019. It is estimated that its value will reach up to \$580 billion by the end of 2025 with a growth rate of 2.7 % between 2020 to 2025. Ovens are operated with the help of LPG/diesel/electricity. The hybrid ovens can give higher energy efficiency along with targeted quality parameters. The previous data such as oven temperature, heat supplied per unit time, the moisture content and variety of the raw material used can be provided to a specific computer programme. The heat recovering unit is a unit which captures the escaping hot air and the same may be reused by pumping the hot air to the front of the oven to preheat the incoming cold air. The biogas can be generated from the bakery waste. There will be 45 % or more methane production from bread and biscuit waste in the retention time of about 12 to 55 days in a biogas plant. To make the bakery fully operated on green technology, solar renewable energy can also be utilized.

Keywords: oven, biogas, solar energy, bakery, green technology, hybrid, temperature, preheat

1.1 Bakery as An Industry

Baked products have been consumed by human beings since 30000 B.C. (Paleolithic period). The reason behind the use of baked products in the daily diet is its taste and fulfillment of daily nutritional requirements. The global market for baked products has reached up to \$495 billion in the year 2019. It is estimated that its value will reach up to \$580.8 billion by the end of 2025 with a growth rate of CAGR of 2.7 % between 2020 to 2025. Globally the baked products largely consumed are bread, doughnuts, cakes, pies and pastries. To prepare such products a wide range of raw materials are brought into use are flour, sugar, starch, dry-fruits, emulsifiers, flavoring agents, preservatives, etc. The major raw material comes from farming. Lot of employment is generated

*Corresponding Author

at farmers' level for production of the raw material. There are a number of businesses which are dependent on the bakery industry. Whereas baking itself provides major employment directly.

1.2 Market for the Baked Products

The biggest market for baked products is Europe. In Europe, Germany consumes the largest amount of bread and rolls. The fastest growth of the baked products market is China and Brazil. These two countries showed 10 % growth rate during the last four years. Next to Europe, the USA is the biggest market and contributes 20 % of the global bakery market. After the USA, China comes second with a 7 % global bakery market.

The Indian bakery industry will touch INR 483 billion in the coming five years. The processed food industry of India comprises cereal, oilseed, pulses, fruit and vegetable processing. Among these, the bakery acquires the biggest section. The country produces abundant raw material required for the bakery industry.

1.3 Machinery Used in an Average Bakery Industry

Common production stages in baking include; mixing, shaping, baking, cooling, and packaging. Certain products require additional production steps. Yeast based products require time for yeast to develop necessitating fermentation and proofing stages. Some bread and roll products are sliced before packaging. Additionally, some products require finishing work, adding decorative items, coatings, etc, after baking and cooling. To prepare different baked products different machineries are required.

The following machinery occupies a bakery unit; Shifter, Mixture/kneaders, Spiral mixture, planetary mixture, Sugar grinder, Dough molder, Proofer, Ovens, Bread slicer, Packaging machines, etc.

Fig.1: Bakery Machinery

1.4 Energy Utilization of Different Bakery Machinery

About 73% of the total energy consumed in bread production is by baking process alone. All the machineries are power operated. Most of the machines are operated with electrical power whereas ovens are operated with the help of LPG gas/diesel or electricity. The electrically operated ovens have their own limitations. Most of the ovens need gas as a fuel. And ovens are the most energy consuming device among the other bakery devices. The profit margin of the bakery totally depends on the efficiency of the ovens. The ovens not only consumes a higher amount of energy but also emits greenhouse gases.

If the efficiency of the ovens is enhanced, the judicious use of the energy will be possible. There are several ways through which energy consumption could be reduced.

1.5 Minimizing Energy use in Baking

If the efficiency of the ovens is enhanced, the judicious use of the energy will be possible. There are several ways through which energy consumption could be reduced. Ways to minimize energy consumption in baking are as follows.

A modern oven utilizes energy more efficiently. To enhance efficiency, the older ovens may be replaced by modern ovens. Faster and better bake uses less energy. So the faster baking may be followed to minimize the energy consumption.

STIR coating changes the wavelength of heat waves (infrared waves). The altered wavelength heat waves penetrate deeper into the product yielding more product volume and faster baking. Therefore, the STIR technology can be adopted to reduce energy use.

Often ovens are heated too much. While preheating the manually operated ovens are overheated. Here negligence of the operator may be the reason.

Low velocity impingement ovens send direct heat straight to the product. In these ovens both the principles of heating are used viz. impingement and convection. In this technology there is an ability of focusing the heat energy on the product instead of the environment which in turn gives shorter baking time.

The hybrid ovens can give higher energy efficiency along with targeted quality parameters and the convection ovens also can provide better quality of the baking. Therefore the direct heater burner type oven technology and convection heating type oven can be combined into a hybrid oven which reduces heat loss and enhances the product quality.

Sometimes the dark coloration (over coloration) issue appears while achieving complete (full) baking. The hybrid baking system addresses such an issue of dark coloration of the baked products. The process of dark coloration starts when the last stresses of moisture from the centre of the product are planned to be removed. Discoloration/dark coloration of the baked product is due to prolonged baking and the prolonged baking means use of additional heat. To avoid discoloration of the baked products, another technique of radio frequency can be opted. Therefore the remaining moisture can be removed by the application of radio frequency.

If a task of even heat distribution is achieved inside of the whole baking chamber, that will result in better baking quality.

The earlier version of the ovens was dependent on a single big tower burner, the present day's ovens come with multiple towers which contribute in better heat distribution hence better baking within a small possible time. The multiple towers helps in operating the ovens efficiently when it is to be run at reduced capacity on the basis of the demand of the particular item. So the burner near to the product may only be operated whereas remaining burners can be put off.

Lowering the flash heat of the oven can be reduced so as to reduce the energy and along with that the baking quality gets improved too.

Thc cmisshild nano emissive oven technology broadens the heat spectrum by radiating infrared heat produced by a gas burner. It gives better heat distribution and reduces energy costs.

The above discussed ways of reduction of energy loss are obtained on either alteration of ovens or adopting correction in manual errors.

1.6 Application of Software for Reduction in Energy Loss

The manual errors could be arrested due to the use of computer programmes. Therefore, the use of software systems in the reduction of the energy loss can be explored.

The previous data such as oven temperature, heat supplied per unit time, the moisture content of the raw material, variety of the raw material used, etc. can be provided to a specific computer programme. Through a computer programme, the oven's control could be operated. The software control system controls over or under heating of the baking chamber and baked products. Hence proper energy utilization will be possible. The software also helps in proper heating of these products as per its stage of baking.

Through the software, the tracking of the product in the oven will also be possible. According to the position of the product in the oven, the programmed burners can automatically time its heat. Some companies provide software based heating software which allows a zone based heating system. And the operator can set temperature based on the heating zone so that the temperature will be the same throughout the particular zone.

Sometimes a gap is found between the products, the software which regulates the burner to its modulation by itself and on arrival of the next product to be baked the burners once again automatically get ignited. The software based control system provides both zone level temperature control system and energy management system which will be available to the operator on his computer screen. The software based system works on tight control of the baking profile of the product. And the oven also gives the predetermined tailor made heat profile based on the product to be baked. The temperature curve of the baking chamber is followed by the oven heating system which helps to obtain a best quality product in relatively little fuel.

There are few continuous type ovens having a multi tower system. Here each tower controls the heat of a particular section and each section has its own temperature zone. Means we can obtain multiple temperature zones based on the raw materials' baking stage. These parameters are governed by PLC, making them operator friendly. The indirect fired tunnel oven expands Bakers oven flexibility with its target products, steaming can be found with these thermador type ovens. This software helps in reduction in energy utilization and also improves the quality of the product.

A real time utility consumption data helps the operator in energy management. The information of gas, power, water, steam and air will be made available to the operator; on the basis of such information the operator's action will be decided.

A fuel flow meter is also provided with some ovens which help the operator to estimate the amount of heat in BTU per kilogram of the baked product is required. And the information works as a baseline for the future fuel demand. Some firms design the oven and analyze the oven control in respect of energy requirements. Along with heating there is an option of cooling the oven as per the product's requirement.

1.7 Utilization Heat from Exhaust Gases of Ovens

To enhance the efficiency of the oven the heat loss must be reduced and the exhaust hot gases can be reused. The heat loss from the oven is due to the poor insulation and idling the oven during the batch of product is changed. While exchanging the racks the heat loss is huge. Some of the manufacturers have adopted heat-recovery units.

The heat recovering unit is a unit which captures the escaping hot air and the same may be reused by pumping the hot air to the front of the oven to preheat the incoming cold air. Sometimes, the air can also be used as a proofreader. To do so, the heat exchanger needs to be used after the chimney. The heat going out from the chimney can be used to heat water or to enhance the temperature of the baking unit during winter. The pre-heating of the cold air saves around 15 kg of gas which is to be sent to the product to be dried into the baking chamber.

Good insulation reduces heat loss from the bakery, reducing the overall ventilation requirements of the building. While changing the racks the ovens kept open which allows heat to escape. To avoid such a heat loss, the number of openings on the ovens can be reduced.

Thermogate feature could be improved so that when the door to the oven is opened, energy loss is minimized.

All these majors enable use of a falling baking profile and better batch-to-batch baking. Some firms reuse heat from the ovens throughout the bakery. They can easily reclaim the heat from combustion through the chimneys and re-purpose the energy for hot water or room temperature inside the bakery.

Exhaust systems also impact how well an oven can maintain its temperature. Managing the oven exhaust becomes a significant consideration for efficiency because too much humidity and moisture in the baking chamber actually dampens the burner output.

While combating the humid environment, some oven process control tools can measure the exhaust flow in real-time and weigh the results against optimal exhaust for that product. And the operator can make adjustments to prevent the oven.

So energy efficiency is the big problem of the entire food processing industry including bakeries. If the green energy concept is adopted, the sustainable bakery business will be possible. The sustainability includes reduction in energy and water consumption and lowering food waste.

Out of the food manufacturing factory's overall annual energy usage, the heat consuming processes, particularly baking, roasting and drying, on average accounts for between 4-10% of energy consumption. It is observed that, most of these processes are over-engineered in which the efficiency is compromised.

The U.S. baking industry consumes over $800 million fuels and electricity per year. To increase predictable earnings, energy efficiency improvement is an important way.

To keep companies in competition, effective ways to manage energy keep costs down. And it is estimated that a well-run energy program can reduce energy costs by 3% to 10%. Improving energy efficiency will help in reducing waste and emissions. The emission causes environmental pollution which is against the green energy.

In energy efficiency, plant-wide energy management practices are combined with energy-efficient technologies. That offers additional benefits of product quality improvement, increased production, and increased overall process efficiency. Once all these are achieved the productivity will also increase.

While making green companies, the overall environmental strategy, energy efficiency improvements will help in reductions in greenhouse gases and other air pollutants.

A technique called heating, ventilation, and air conditioning (HVAC), provides fresh filtered air, heating, cooling, and humidity control in the building. The main goal of this technique is to provide indoor air quality. The quality will depend on many factors, such as thermal regulation, control of internal and external sources of pollutants, supply of acceptable air, removal of unacceptable air, occupant's activities and preferences. Here, Air exchange between outdoor and indoor air is required to maintain good air quality and bearable indoor temperatures.

The exchange of air depends on the ventilation process and infiltration process. Ventilation may be done naturally or through mechanical forced ways. The mechanical system includes, using fans or vents.

The Infiltration of air is the flow of outside air into the building through cracks and other unintentional openings. That can be minimized. Sizing equipment properly and designing energy efficiency into a new facility generally minimizes the energy consumption. The long-term money saving is possible after applying such a technique. If the energy-efficient HVAC equipment is installed during construction itself the cost of installation remains low compared to upgrading in the existing building. The initial commissioning of the building is essential where it is to be ensured the energy performance and operational goals are met.

Green technology in the form of solar air heating systems could be adopted. A solar panel is introduced on the building which absorbs the solar radiation. The fresh air enters beneath the panel where it is heated as it passes over the warm absorber, and fans distribute the air. This measure is best applied at buildings in cold climates. Use of a reflective coating on the roof of buildings in sunny, hot climates can save on air conditioning costs inside.

1.8 Utilization of Bakery Waste to Produce Biogas

As far as green baking technology is concerned the waste from the bakery industry needs to be utilized further to produce biogas and manure. Bakery waste contains easy to disintegrate sugar and supporting nutrients. The biogas can be generated from the bakery waste. There will be 45 % or more methane production from bread and biscuit waste in the retention time of about 12 to 55 days in a biogas plant. All the bakery waste gives pH value in the range of 5.3 to 7.4. While in biogas plant the required temperature for digestion of these wastes will be between 20 °C to 40 °C (mesophilic conditions). The slurry output of the biogas plant can further be utilized as manure in the farmers' field.

Fig. 2: Biogas plant

Fig. 3: Solar panel

The required modifications in the present ovens may be done so that the same biogas can be further used as a fuel for the ovens. To make the bakery fully operated on green technology, solar renewable energy can also be utilized. The bakery room and the machines control panels are operated with the help of electrical power. If solar energy is harvested through solar panels and generated electricity is used for operating lights, fans and electrical control panels of the various machinery then the true green technology based bakery is possible. Solar thermal based systems for bakery ovens can be employed.

1.9 References

Ibisworld, market research report global bakery goods manufacturing industry, https://www.ibisworld.com, 2020.

Indian Institute of Solar Energy, solar panel, https://indian- institute-of-solar-energy.business.site/,2020.

Khoshkhoo RH, Omrani MM. Energy Audits and Recovery in the Production of Industrial Bread and Pastry. 3rd Conference on Advances in Mechanical Engineering, 156-171, December 2017.

Kodag V, Kulkarni GS. A review on biogas production from bakery waste. International Journal of Trend in Scientific Research and Development, 3, 2, 2333-2337, 2018

Kot W, Adamski M, Durczak K. usefulness of the bakery industry west for biogas production. Poland Source Journal of Research and Applications in Agricultural Engineering, 60, 2, 2015.

Masanet E, Therkelsen P, Woeewl E. Energy efficiency improvement and cost saving opportunities for the baking industry. Environmental Energy Technologies Division. Ernest Orlando Lawrence Berkeley National Laboratory, 3-8, 2012.

Mukherjee S, Asthana A, Howarth M , Mcneill R, Frisby B. Achieving Operational Excellence for Industrial Baking Ovens. 2nd International Conference on Sustainable Energy and Resource Use in Food Chains, ICSEF. Energy procedia. Science Direct, 17-19 October, 2018.

Orange foodstuff, bakery machinery, www.orangemultiventures.in.2019

Paton JB, Energy utilization in commercial bread baking. Unpublished Ph.D. thesis. The University of Leeds, 15-24, 2013.

Researchgate, Energy Audits and Recovery in the Production of Industrial Bread and Pasty, https://www.researchgate.net/publication/322147648, 2017.

Swachhindia, Haryana Government, agricultural waste, https://swachhindia.ndtv.com/agriculture-waste-haryana-government-inks-pactwith-ioc-to-setup-bio-cngplants-25674/, 2020.

Vannoni C, Schweiger H. Energy audit summary report audit No. 72. Food industry bakery, Einstein audit summary report. 8-10, 2012.

2

Ozone Treatment: The Green Technology in Food Industry

Krantidip R. Pawar, Vinod T*. Atkari, Dilip A. Pawar* and *Pravin D. Ukey

ABSTRACT

The food industry has many the environmental challenges in terms of emerging microbial strains, bacteria, viruses and accumulation of toxic chemicals. Ozone (O_3), environmental bluish gas is an extremely potent oxidant, which is effective solution for disinfection, sanitization for various kinds of products. The application of ozone treatment in agro processing has become green technology with increasing acceptance. It can apply in liquid form, gaseous form to sanitize food, food packaging materials, process water and to disinfect equipments. Ozone can be generated by different methods such as ultraviolet radiation method, corona discharge method etc. In this chapter, ozone properties, its generation, antimicrobial power and advantages with application of ozone has covered. Fruits and vegetables treatment with ozone have shown increased shelf-life and many hazardless advantages of the products. Some products preserved with ozone. Anti microbial activity, no residues in foods, assurance about preservation of sensory, nutritional and physicochemical properties of food have made ozone application as a promising technology for all kinds of food products

Keywords: Ozone, Disinfectant, Sanitation, Antimicrobial, Chlorinated water, Ozone generation

2.1 Introduction

Now days, pathogens such as bacteria, viruses and many other microbes on fruits, vegetables and their products are a primary issues related to food-safety. Many thermal unit operations and chemical treatments are applied to minimize these pathogenic microorganisms. However, these technologies, affect the quality of foods. Certain novel techniques like high pressure processing,

*Corresponding Author

pulse electric field and many others prevent the food quality losses to certain extents. Sanitization, aseptic processing, and maintaining hygiene are the key parts to get hazard free safe food products. Generally, sanitization is done with chemical treatments, thermal treatments, which are crucial to maintaining the quality; also it has many drawbacks like residues retentions, nutritional losses and other quality losses. Therefore food industry is in need of green processing technologies in order to meet the consumer expectations. The ozone application in food sector is one of the promising technologies to minimize these kinds of drawbacks. Ozone has shown strong and rapid antimicrobial action against all kinds of spores, fecal and pathogenic microorganisms, and viruses, as compared to chlorine. Ozone use has many advantages and applications, like no chemical residues and natural degradation into oxygen. (Khadre *et al.,* 2001).

Many food disinfectants are available in the market and commercially applied in food sectors like chlorine, Iodophors, Quaternary Ammonium Compounds etc. Every disinfectant has certain disadvantages (Table 1.)

Table 1: Disadvantages of Some Widely used Chemical Disinfectants

Disadvantages of some disinfectants	Chlorine and its compounds	Iodophors	Quaternary Ammonium Compounds
	• Organic matter causes a quick reduction in bactericidal effectiveness. • Effectiveness reduces as pH increases. • Highly dissipates in hot water. • Corrosion to metal . • Affect to skin if contact. • Low shelf life. • Off odor.	• Slow reaction at pH 7.0 above, • Quick vaporizes at 120 °F. • Less effective against bacterial spores than hypochlorites • Stain some plastics and porous surface • Highly expensive	• Not compatible with hard water and most detergents. • Forms film. • Produces foam in mechanical operations. • Selective in destruction or inhibition of various types of organisms. • Requires higher concentration for better action than chlorine or iodine. • Expensive.

Ozone has a strong oxidant and strong antimicrobial components which destroys 99.9% pest and microorganisms due to its potential oxidizing capacity. Due to oxidation process any pathogens or contaminants can be disinfected or removed. Ozone has shown to be effective over a much wider spectrum effect on the microorganisms than chemicals disinfectants. It is the strongest disinfectant of all, as it applies in water treatment. (Khadre *et al.,* 2001; Kim *et al.,* 1999).

In France on year 1886, first time Ozone has been adopted for production of potable water. In ozone treatment, foods to be decontaminated is exposed aqueous and/or gaseous phases of ozone at a constant flow per unit time, constant pressure and particular concentration based on the percentage of contamination. There are many applications such as surface hygiene, plant sanitization, equipment sanitization, waste water treatment, controlling level of chemical oxygen demand (COD) and biological oxygen demand (BOD) (Brooks *et al.,* 1990). Ozone has been used for disinfect the water and remove off odor (Rice, 1999; Muthukumarappan *et al.*, 2000). Utilization of ozone in member countries for a long time has reported by many authors (Guzel-Seydim *et al.,* 2004). In US ozone treatment in the food industry has not been widely used till 1982. In 1982 first time for use of ozone in bottled water the federal agency US FDA approved generally recognized as safe (GRAS) status. Ozone has also been declared as GRAS for use in food processing by the US FDA in 1997 and allowed as an antimicrobial food additive in 2001 (Graham, 1997).

2.2 Ozone Structure and Property

Ozone (O3) is naturally present in the atmosphere in a gaseous form. Lightning or high energy UV radiation responsible for creation of natural Ozone in the atmosphere (Jakob and Hansen, 2005). Ozone is an allotropic form of O2 available in different form and mostly are in chain arrangement. It have molecular weight of 48.

2.3 Ozone: Antimicrobial Action

Ozone has antimicrobial action against bacteria (vegetative and spore), fungi, viruses and other similar microbe (Ligimol *et al.,* 2002). Inactivation of Microorganism by ozone treatment is a complex process. Ozone reacts with microbial constituents like cell walls, cell membranes, endosperm coats, the cytoplasm, viral envelopes and virus capsids (Khadre *et al.,* 2001).

Bacteria - Ozone inactivates both kinds of bacteria. Gram-positive bacteria were less sensitive to ozone than Gram-negative organisms. Ozone at different doses and exposure times showed *E. coli* inactivation *(*Finch *et al).* High concentrations of airborne ozone have achieved 99% death rates for *E. coli* and S. *aureus* (Kowalski *et al.,* 1998). Moore *et al.* observed inhibitory effect of ozone on *E. coli, Staphylococcus aureus, Serratia liquefa ciells, Listeria innocua* and *Rhodotorula rubra.* Dose of 1.5 ppm ozone in distilled water shown decrease of *Salmonella Enteritidis* population was by 6 log units (Kim *et al.,* 1999).

Fungi - *Candida albicans* and *Zygosaccharomyces bacilli* have considerable reduction by ozoneated water treatment (Restaino *et al.,*1995*)* Kawamura *et al.* found yeasts on ozone treatment shown more sensitive than molds.

Whereas, in the case of *Aspergillus niger* spores, less reduction was noticed. Many organism found inhibitory and Moore *et al.* found that yeast species were more resistant than bacteria towards ozone activity. Beuchat *et al.* (1999) worked and reported the susceptibility of conidia of aflatoxigenic aspergilli to ozone.

Viruses – Naturally ozone acts as a virucidal agent inactivates viruses on short exposure. Viruses with cells are high resistant to ozone compared to purified virus II. Herbold *et al.* have tested flowing water in steady having pH 7 at 20°C to find resistance of viruses and bacteria to ozone. When the ozone demand of the medium is low then low concentration levels ozone and less process times are generally adequate for inactivation of viruses whereas when the ozone demand of the medium is high (e.g., in wastewater), long contact time and high ozone concentration are sufficient to inactivate viruses (Kim *et al.,* 1999).

Protozoa - Ozone is more effective than chlorine against *Cryptosporidium* and *Giardia. Naegleria gruberi* cysts were more resistant than *Giardia muris* to ozone. Korich *et al.* reported that more than 90% reduction in the population of intestinal parasite, *Cryptospordiumparvwn,* within 1 min in ozone demand free water. Ozone is more effective against protozoan parasites in water systems as compare to other chemical disinfectant like chlorine or chlorine dioxide.

2.4 Ozone Production

Many methods are adopted for production of ozone. The corona discharge method uses high-energy electrical field, photochemical method based on ultraviolet radiation and chemical method where oxygen molecules converts into ozone for continuous use. In which photochemical and corona discharge are used commercially (Guzel-Seydim *et al.,* 2004). Ozone production can be done by electrolysis method, phosphorus reactions with water, and radiochemical procedure. However these procedures are in their in the early hours stages of development and not much cost effective.

2.5 Equipment for Ozone Treatment

Generally, ozone may utilize in any aqueous or gaseous phases or in combination of both. Ozone treatment apparatus mainly consists of the gas/pure air, an ozone generator system, an electric power supplier, contactor (water phase ozone), reactor, surplus gas elimination system, and ozone analyzer (Bablon *et al.,*1991). Ozone treatment in the water phase requires similar elements with additional sample pH controller unit (Naito and Sawairi, 2000).

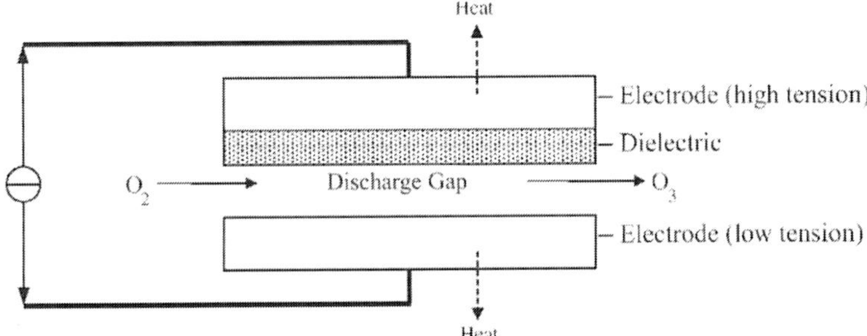

Fig.1: Schematic diagram of Corona Discharge method for ozone generation (Rice *et al.*,1981)

2.6 General Advantages of Ozone Treatment

- Any hazardous chemical residue frees the process.
- Better fumigation than conventional fumigation method.
- Ozone treatment is versatile to all kinds of agro produce foods, meat and sea foods, value added products.
- No adverse effect on texture of fresh & frozen foods.
- Effective in lowering the microbial load leads to enhanced shelf life of produce.
- No adverse effect on sensory and visual quality of the product.
- Non toxic
- Environment friendly and no pollution after treatment (Greene *et al.*, 2012).

2.7 Disadvantages of Ozone Treatment

- Microorganisms sensitivity to ozone differ by several factors like type of product, kind of microorganism, level of contamination, physiological state of the bacterial cells, the physical state of ozone, kind of an organic material.
- Different microorganisms have different resistivity to ozoneconcentration; therefore effectiveness of dose for particular microorganism is different. (Restaino *et al.*, 1995).
- Higher ozone concentrations may negatively affect food quality preservation.
- High ozone treatment and retaining time may cause reduction of vitamin, polyphenol, and volatile compound.

- Ozone resistant compounds present in water may cause partial oxidation (Hoign-e, 1998).
- Ozone is unstable in water cause improper ozone concentration doses.
- Ozone treatment cannot be economically feasible to small manufacturer.
- Unawareness about ozone treatment leads to poor acceptability by manufacture and consumers.

2.8 Ozone Application in Food Industry

Ozone technology becomes useful during various unit operations specially used for washing equipment and packaging (Graham 1997). Ozone treatment is widely applied in many food sector. However, an application of ozone use widely depends on the nature, composition, structure of food surface, kind of microbe, end use of product and the degree of final acceptance of microorganisms with food.

A. Fruits & Vegetables Processing

Farm fresh fruits and vegetables are highly prone to contamination by environmental microorganism since from pre harvesting stage. Sivapalasingam *et al, 2004* reported increased food borne disease outbreak linked with minimal processed fruits and vegetable during past decade. Many chemical sanitizers & disinfectants are applied to remove surface contaminations, which leads to safety issues due to residues remain over it. Food industries are now moving to use ozonated water for washing and cleaning agro produces. Kim *et al.* have used ozonated water to wash shredded lettuce and obtained 2- log cfu/g reduction in total plate counts. Ozone efficacy against microbe on surface depends on ozone delivery method. Washing apple with bubbling ozone inactivate *E. coli* O157:H7 on apple surfaces in pre-ozonated water was reported by Achen and Yousef (2001). Ozone treatment on Table grapes increased shelf life due to deactivation of fungi (Sarig *et al.,* 1996).

Fruit and vegetable shelf life mainly controlled by ethylene content produced during respiration and ripening. High ethylene production leads to more ripening and senescence in fruits and vegetable. Ozone treatment is effective method to control the ripening and senescence by removing ethylene through chemical reaction leads to extended shelf life of fruits & vegetables. The shelf-life of potatoes would be enhanced upto 6 months at temperature range from 6 to 14°C and Relative humidity 93 to 97 per cent with 3 ppm of ozone without affecting the quality. Ozone can also be used for increasing the shelf-life of blackberries, by preventing fungal growth without causing any observed defects, and surface colour of berries for 12 days.

B. Meat and Meat Processing Sector

Pathogenic microorganisms are highly responsible to cause hazardous contamination to meat and meat products during handling, processing, slaughtering and distribution operations (Zhao *et al.*, 2001). During washing of beef carcasses with natural water followed by ozonated water was more effective than that of treatment with trisodium phosphate, acetic acid or a sanitizer. On the contrary, it have been found that treatment with ozone was an improvement over control but it showed no advantage over conventional washing in reducing microorganisms. Carcasses and cut fresh meat decontamination may do with chemical sanitizer but the safety of this method was questionable. Therefore ozone treatment is the best alternative sanitizer against meat and meat product contamination. Many researchers confirmed the applicability of ozone as safe disinfectant for meat industry. Meat and meat product contaminated with faecal materials which may includes pathogens due to improper good handling practices. These faecal contamination and pathogen may removed by antimicrobial ozone treatment.

Greer and Jone studied on impact of ozone application on beef carcass bacterial spoilage, meat quality and carcass trexture. They reported that under ozone atmosphere psychrotrophic bacterial growth was retarded on carcass surfaces. (Greer and Jones, 1989).

C. Dairy Sector

Milk handling equipments and machineries adhered with milk residues and pathogenic bacteria during milk handling and processing which are rigid to remove. Heavy microbial load and milk residues on handling equipments lead to deterioration in quality of milk. Ozone treatment is best substitute solution to chemical based sanitizers to remove these residues and bacteria from surfaces (Guzel- Seydim *et al.,* 2004). Milk product like cheese can be prevented by mould growth and air borne moulds during ripening and storage. Shiler *et al* have also suggested a method of ozonation of cheese for ripening and storage without any harm to cheese and its wrapping materials. Swedish company has developed pasteurization process which included pre-ozonation treatment followed by conventional pasteurization (Pastair 2014). The ozone application has claimed to extension of shelf life with better quality. Dried milk product like milk powder also influenced by ozone treatments in chemical, physical, functional and organoleptic properties. Kurtz *et al.*, 1969 reported that skim milk powder have significant sensory score which is prepared under surroundings ozone level of 32 ppb. Also, with wide application of ozone treatment, dairy waste water had effectively treated with ozone treatment to control over pollution in terms of BOD and COD (Laszlo and Jeno, 2016).

D. Grain Industry

Many insects and microbes like mould, fungi, bacterial were grow during storage of grains and which leads to post harvest loss around 3-10 per cent yearly (Jayas 1999). Many techniques are used to reduce the losses but have limitation in terms of adverse effects. Ozone treatment with proper concentration dose will reduce the level of contamination and deterioration of the grains. Vikas Chandra Verma *et al.*, 2018 reported ozone treatment efficiency depends on the kind of nature, processing state and ozonation conditions. Red flour beetle, maize weevil and larvae Indian meal moth had evaluated for effect of ozone fumigation in corn grain. He reported at lower dose mortality (77–99.9%) of insect depends upon the insect species. No effect on germination of grain was found even ozone level was controlled below 0.98 mg/g of grains per min. Germination rate of corn seed had found higher when treated with ozone for short term (Violleau *et al.,* 2008).

Wheat grain treated with ozone earlier than processing was studied to improved flour quality and safety (Yvin *et al.,* 2001). Three days ozone treatment showed 92–100% mortality of Sitophilus zeamais (Motsch.), Tribolium castaneum, adult red flour beetle, larval moth and Plodia interpunctella. Also, the level of contamination of fungus Aspergillus parasiticus Speare on surface was reduced 63%. The additional advanteage of ozone treatment is ozone dispose rapidly to molecular oxygen with no any chemical residue and hence proved as an effective technology for grain protection without affecting its quality. (Pawar *et al.,* 2015).

E. Dry Foods

Dry foods like dehydrated vegetables, spices, powders get contaminated by various pathogens, yeast, mould due to mishandling and packaging. Dry foods are hygroscopic in nature and high potential to contaminate due to increased water activity. Efficacy of ozone treatment on dry food mostly depends on surface properties of products, ozone dose, reaction rate, temperature, humidity in the environment and water activity of the product (Kim *et al.,* 2003). Aflatoxin as most toxic substance is commonly found in dry foods. Detoxification of aflatoxin by ozone treatment has reported (Akbas and Ozdemir, 2006).

Cereal flour and ground pepper had achieved microbicidal effect when treatment with high ozone concentration and long exposure duration (Naitoh *et al.,* 1989). Naitoh *et al.,* have obtained 1 to 3 logs reduction in count of *Bacillus* and *Micrococcus* of cereal grains, peas, beans and spices on treatment with < 50mg/L ozone. In general, high exposure time and low treatment temperature resulted in high microbicidal activity in dry food products. Ozone concentration of < 5 ppm can be used effectively without oxidative changes in lipid containing foods. Ozone decreased essential oil content in some spices, thereby having a negative impact on sensory quali ties. Galdun *et al.* tested the effect of ozone on garlic during long-term cold storage.

F. Packaging Material and Food Contact Surfaces

Packaging materials usually contain many microbial contaminants. Many sterility treatments are utilized to obtain contaminant free packaging materials including hydrogen peroxide (H_2O_2), other sanitizers, thermal treatment like heat, hot air, pressurized steam, Ultra violate radiation (Gardner and Shama, 1998). Sterilization of packaging materials by H_2O_2 has several disadvantages like retention of residue after treatment (Yokoyama, 1990). Ozone applied in gaseous and aqueous states has been explored for surface sterility of packaging materials and tools (Pascual *et al.*, 2007). Bacterial biofilms of *Pseudomonas fluorescence* and dried films of *B. subtilis* spores on surface of multi laminated aseptic food packaging can be tested with help of ozone treatment (Khadre and Yousef, 2001a). Multilaminated packaging sterility was achieved when treated with 5.9 μg/ml aqueous ozone for 1 min. Oak barrels used for aging wine were disinfected when treated with ozone. Spoilage yeast has been proven to control over fungus (Day, 2004). Many studies showed ozone treatment inactivate *P. fluorescence* biofilms on stainless steel.

G. Miscellaneous Applications

Ozone has been proven to best alternative of chemicals for disinfection and sterilization of water used by food industry for processing. Ozone has capability of destroying chlorine by-products, chemicals and organic compounds in the water without leaving any hazardous residues. Depending on the source application of ozone ranges from 0.5 to 5 ppm with less than 5 min contact time. Ozone can also be used to remove many minerals like manganese, iron, sulphur etc. Ozone treatment helps in control taste and odour of water. Ozone treatment ensures continuous availability water with high quality and free from toxic chemicals for the industry. Sander has worked on quality deterioration of fruit juices and liquid dairy products treated with ozone showed positive results. Moore *et al* have recommended ozone as an effective terminal disinfectant, if applied after adequate cleaning. Ozone was found to be effective on airborne microorganisms in different air disinfecting systems.

Ozone treatments on waste water have shown best results in terms of control over BOD, COD and other solid residues. It has been used to disinfect, to remove turbidity, colour, odour and to reduce the organic loads of wastewater. Ozone can be used for recycling of water in other industries, particularly in food sector.

2.9 Conclusions

Sanitization and safety process during food processing become important step that ensures diffusion, inactivation of hidden and entrapped pathogens. Ozone efficiency, retention time and spontaneous decomposition to nontoxic products (i.e., O_2) make this ozone a viable disinfectant for ensuring the safety and quality of food products. Ozone application has wide scope in food industry. In modern years processor and consumer became food process literate and have changed their perception about ozone treated foods.

2.10 References

Achen, M. and Yousef, A.E. 2001. Efficacy of ozone against Escherichia coli O157:H7 on apples. Journal of Food Science, 66, 1380-1384.

Akbas, M. and Ozdemir, M. 2006. Effect of different ozone treatments on aflatoxin degradation and physicochemical properties of pistachios. Journal of the Science of Food and Agriculture. 86, 2099-2104.

Brodowska, A. J., Smigielski, K., Nowak, A., Brodowska, K., Catthoor, R. and Czyzowska, A. 2014. The impact of ozone treatment on changes in biologically active substances of cardamom seeds. J. Food Sci. 79(9): C1649–1655.

Beuchat L R, 1991. Surface disinfection of raw produce, Dairy Food Environ Sanitation, 12,6.

Brooks, G. M. and Pierce, S. W. 1990. Ozone applications for commercial catfish processing. Paper presented at 15th Annual Tropical and Subtropical Fisheries Technological Conference of the Americas, December 2-5, Orlando, Florida.

Day, C. 2004. Developments in the management of Brettannomycess. Wine Industry Journal. 19, 18-19.

Finch G R, Smith D W & Stiles M E, 1988. Dose-response of *E. coli* in ozone demand-free phosphate buffer, Water Res, 22 , 1563.

Galdun T I, Postol A Y & Lukovnikova G A, 1984. Changes in the carbohydrated comples of garlic during long term cold storage with ozonization and ionization treatments, Tovarovedenie, 17, 43.

Gardner, D.W.M., and Shama, G. 1998. The kinetics of Bacillus subtilis spore inactivation on filter paper by u.v. light and u.v. light in combination with hydrogen peroxide. Journal of Applied Microbiology, 84, 633-641.

Guzel-Seydim, Z. B. et al., 2004. Use of ozone in the food industry. Lebensm.-Wiss. u.-Technol. 37, 453–460.

Graham, D. M., 1997. Use of ozone for food processing. Food Technology, 6 (51), 72-75.

Greer, G.G. and Jones, S.D.M. 1989. Effects of ozone on beef carcass shrinkage, muscle quality and bacterial spoilage, Can Inst Food Sci Technol J, 22: 156–60.

Greene A K, Few B K & Serafini J C, 1993. A comparison of ozonation and chlorinating for the disinfection of stainless steel surfaces, J Dairy Science, 76, 3617.

Greer, G.G. and Jones, S.D.M. 1989. Effects of ozone on beef carcass shrinkage, muscle quality and bacterial spoilage, Can Inst Food Sci Technol J, 22: 156–60.

Guzel-Seydim Z B, Greene A K and Seydim A C. 2004. Use of ozone in the food industry. LWT - Food Science and Technology, 37 453–460.

Hoigné, J., and Bader, H. Ozonation of water. 1975. Role of hydroxyl radicals as oxidizing intermediates. Science (Washington, DC, United States) 190, 782-784.

Herbold K, Flehmig B and Botzenhart K, 1989. Comparison of ozone inactivation in flowing water of Hepatitis A virus poliovirus I and indicator organisms, Appl Environ Microbiol, 55, 2949.

Jayas, D.S. 1999. Grain preservation: researchers aim to decrease waste by improving storage techniques, Resource, 7: 7–8.

Jakob SJ and Hansen F. 2005. New Chemical and Biochemical Hurdles. In Emerging technologies for Food Technology. Edited by Sun, D. Elsevier Ltd., 387-418.

Khadre M A, Yousef A E and Kim J G, 2001. Microbiological aspects of ozone applications in food: a review. Journal of Food Science, 66, 1242–1252.

Kim, J.G., Yousef, A.E., and Chism, G.W. 1999. Use of ozone to inactivate microorganisms on lettuce. Journal of Food Safety, 19, 17-34.

Kells, S.A., Mason, L.J., Maier, D.E. and Woloshuk, C.P. 2001. Efficacy and fumigation characteristics of ozone in stored maize, J of Stored Products Research, 37: 371–82.

Kowalski W J, Bahnfleth W P and Whittam T S, 1998. Bactericidal effects of high air-borne ozone concentrations on E. coli and Staphylococcus aureus, Ozone Sci Engineering, 20, 205.

Kurtz F E, Tamsma A, Selman R L and Pallansch M J. 1969. Effect of pollution of air with ozone on flavor of spray-dried milks, Journal of Dairy Science, 52 158–161.

Ligimol James, A K Puniya, V Mishra and Kishan Singh. 2002. Journal of Scientific & Industrial Research,- Vol. 61, 504-509.

Laszlo Varga and Jeno Szigeti, 2016. Use of ozone in the dairy industry: A review.International Journal of Dairy Technology, 69(2), 157-168.

Mahapatra, A., Muthukumarappan, K., and Julson, J. 2005. Applications of ozone, bacteriocins, and irradiation in food processing: A review. Critical Reviews in Food Science and Nutrition. 45, 447- 461.

Moore G, Griffith C and Peters A. 2000. Bactericidal properties of ozone and its potential application as a terminal disinfectant. Journal of Food Protection, 63, 1100–1106.

Patil S and Bourke P. Ozone processing of fluid foods. 2012. In Novel Thermal and Non-Thermal Technologies for Fluid Foods, 225–261.

Pastair, Cold pasteurization: what could be more natural? [Internet document] URL http://qb.se/cmarter/files/20120330-OZ7Z-4PBZVCSC.PDF. Accessed 07/12/2014

Pascual, A., Llorca, I., and Canut, A. 2007. Use of ozone in food industries for reducing the environmental impact of cleaning and disinfection activities. Trends in Food Science and Technology, 18, S29-S35.

Pawar S.G., Pardeshi I.L., Bajad V.V., Surpam T.B. and Rokde H.N. Ozone: 2015. A New Controlled Strategy for Stored Grain, Journal of Grain Processing and Storage, Vol 2.1, 01-10.

Restaino L, Frampton E W, Hemphill J B & Palnikar P, 1995. Efficacy of ozonated water against various food-related microorganisms, Appl Environ Microbiol, 61, 3471.

Rice, R. G., Robson, C. M., Miller, G. W., & Hill, A. G. 1981. Uses of ozone in drinking water treatment. Journal of the American Water Works Association, 73(1), 44–57.

Rojek U Hill A & Griffiths M, 1995. Preservation of milk by hyperbaric ozone processing, J Dairy Sci, 78, 125.

Shiler G G, Eliseeva N N and Chebotarev L N. 1978. Use of ozone and ultra-violet radiation for the inactivation of mould spores. Proceedings of the 20th International Dairy Congress, E 616.

Selma, M.V., Beltran, D., Allende, A., Chacon-Vera, E., and Gil, M.I. 2007. Elimination by ozone of Shigella sonnei in shredded lettuce and water. Food Microbiology, 24, 492-499.

Sivapalasingam, S., Friedman, C.R., Cohen, L., and Tauxe, R.V. 2004. Fresh produce: a growing cause of outbreaks of foodborne illness in the United States, Journal of Food Protection, 67, 2342-2353.

Sarig, P., Zahavi, T., Zutkhi, Y., Yannai, S., Lisker, N., and Ben-Arie, R. 1996. Ozone for control of post- harvest decay of table grapes caused by Rhizopus stolonifer. Physiological and Molecular Plant Pathology , 48, 403-415.

Verma V. C., 2018. Applications and Investigations of Ozone in Cereal Grain Storage and Processing: Benefits and Potential Drawbacks, International Journal of Current Microbiology and Applied Sciences Special Issue-7: 5034-5041.

Violleau, F., Hadjeba, K., Albet, J., Cazalis, R., and Surel, O. 2008. Effect of oxidative treatment on corn seed germination kinetics, Ozone, Sci and Engineering, 30: 418–22.

Yokoyama, M. 1990. Aseptic packaged foods. In Food Packaging. Kadoya, T. (ed) Chapter 12, p.213-228. Academic Press Inc., New York.

Zhao, C., Ge, B., De Villena, J., Sudler, R., Yeh, E., Zhao, S., White, D.G., Wagner, D. and Meng, J. 2001. Prevalence of Campylobacter spp., Escherichia coli, and Salmonella serovars in retail chicken , turkey , pork, and beef from the Greater Washington, D.C. area. Applied and Environmental Microbiology, 67, 5431-5436, 2001.

3

Application of Green and Eco-Friendly Techniques in Food Packaging

Amit Pratap Singh, Anurag Singh* **and** ***Ashish Khare***

ABSTRACT

Packaging plays an important role in food industry. The huge waste of packaging generated every year poses a great threat to the environment. The recycling and reuse of the food packaging is limited. Plastic packaging in particular are the serious concern with respect to their disposal. This is the high time when eco-friendly green technologies for the food packaging should be explored. This chapter discusses various possibilities to use the green and eco-friendly approaches for food packaging.

Keywords: Green packaging, bio-degradable, recycle, reuse, edible packaging.

3.1 Introduction

Food Packaging has an important role in the preservation of quality and safety, and also extension in shelf life of a food product in the supply chain (Ambrose, 2020; Williams *et al.*, 2012). Food packaging is a technique that helps a fresh or finished product in reaching to the ultimate consumer in sound and safe condition from the production centre at reasonable cost (John, 2013). Plastic has been used as packaging material since very long. Such polymeric materials are manufactured by the use of derivatives of petroleum and hence are costly and also pollute the environment as these decomposes in many years (Guillard *et al.*, 2018). Dumping of plastic wastes results in leaching of hazardous chemicals in the soil that ultimately affects the soil fertility, while on incineration toxic gases emits that causes harm to the environment. Due to the presence of colour and other additives, recycling of plastic based waste cannot be considered as effective solution of this problem (Barlow and Morgan, 2013). Therefore, there is a dire need of finding possibilities for the green and eco-friendly techniques in the area of food packaging.

*Corresponding Author

3.2 Functions of Food Packaging

From the consumers view point, functions of food packaging are - Preservation, Protection and Presentation. However, these functions have least relevance with industrial products (Mesias *et al.*, 2015). Two basic objectives are served by packaging are marketing and logistics. As per John (2013), the marketing functions are:

a) Gives information to customers about the final product

b) Helps in promotion of product with attractive printing and graphics

c) Communicate with consumer

d) Act as the interface between consumers and company (silent salesman)

The logistic functions are - containment, protection, dispensation, unitization, convenience and communication. In an organization, the packaging functions are interrelated with different departments like production, legal, R & D, purchase, marketing and finance. That is why, packaging department is considered as the nucleus in any fast moving consumer goods (FMCG) industry (Pongracz, 2007).

3.3 Food Packaging Materials

Packaging materials for foods are classified into two main categories, these are primary packaging materials and ancillary packaging materials. Primary packaging materials are further divided into three groups: flexible packaging materials, rigid packaging materials and semi-rigid packaging materials. Flexible packaging materials include paper, paper board, cellophane, aluminum foil, jute or hessian materials, plastic woven sack, plastic films and laminates. Rigid packaging materials include metal containers, glass containers, plastic containers, plastic crates, wooden containers, corrugated fiberboard boxes, fiber drum, etc. Semi-rigid packaging materials include aluminum collapsible tube, plastic collapsible tube, composite container and paper based carton. Ancillary packaging materials are printing ink, adhesives, labels, cushioning materials, straps, tapes, nails, hook, clips, etc. (John, 2013).

3.4 Glass

Glass is an ancient packaging material and withstand the challenges with different packaging materials and is continued till date. Glass containershave wide applications in the packaging of pharmaceuticals, dairy, liquor, breweries, food, soft drinks, cosmetics, chemicals, inks and other industries. Glass possesses certain unique properties that have made it to lead over otherpackaging materials. These properties are chemically inert, non permeable, transparent, moldable, strength, light weight and unlimited supply (Pongracz,2007).

3.5 Metal Cans

Metal cans are generally made either from aluminum, tin plate or tin free steel. In food packaging, tin plate and aluminum containers are extensively used. Tin plate containers are the most popular metal containers and has been constantly used in food packaging for more than 50 years (Deshwal and Panjagari, 2020). Metal packaging is considered the oldest one for food packaging industries after glass containers. Tin plate containers are of two types - open top containers and general lined container. Open top containers are generally 3-piece can. Now Drawn and Wall Iron can has also been developed that are called two piece cans. DWI cans are commonly used for pressurized beverages and beer. General lined containers are mainly used for packing of bakery products, hydrogenated oils and confectionery items (Pongracz, 2007).

3.6 Aluminum Foil

Aluminum foil is the best flexible packaging material in terms of barrier properties against moisture, gases, light, aroma, etc. Not any other flexible material can compete the properties of aluminum foil as it retains all the metallic properties of aluminum. Important properties of aluminum foil are impermeability, non toxic, stable, light & heat barrier, tasteless & odourless, tagger and retortable. It can be used for decorative label, confectionery, biscuitwrappers, milk products, multilayer laminates, standup pouches, tea chest lining, etc. (John, 2013).

3.7 Plastic Materials

A plastic material is solid at ordinary temperatures and on the application of pressure and heat, allows considerable and permanent change of form without loss in its coherence property. Plastic materials are the most versatile group ofmaterials used in food packaging. In India, due to easy availability of resins, use of plastic is growing fast (Dey *et al.*, 2020). Plastics are light weight, strong, hygienic, non- conductive and do not rust, rot or react with most of the chemicals. The plastic materials used for packaging include the polyolefin, principally polyethylene, and polypropylene, poly vinyl chloride, polystyrene and polyethylene terephthalate (John, 2013).

3.8 Impact of Food Packaging Materials on Environment

Food packaging materials have substantial role in environment degradation. Some packaging materials such as plastics are dumped almost everywhere that cause an adverse effect to the ecosystem and pose a significant threat to the environment (Kooijman, 2000). Some of the packaging materials are manufactured by using natural resources and therefore destroying the animal's habitats.

Forest resources are commonly used for packaging materials such as wood and paper. Continuous deforestation threaten its growth balance. This threatened growth balance result in soil erosion, desertification, shortage of water and a series of ecological problems (Pongracz, 2007). The process of manufacturing the packaging materials also degrades the environment due to the production of toxic gasses into the air and cause the air pollution. The packaging wastes that are dumped irresponsibly cause the environmental pollution as most of the wastes do not degrade. Non degradable packaging materials stay behind for an immensely longer time. As a result, these destroy the soil and other components of the environment and aquatic systems. Packaging waste that are carefree dumped may be ingested by animals may cause death of animals. Waste dumped in the ocean, may disturb the marine environment and may cause death of aquatic animals (Bugusu, 2007).

3.9 Green Technology

Green technology is the application of green chemistry and environmental science and also electronic instruments for environmental monitoring in conservation of the environment and natural resources and restricting the adverse consequences of human intervention (Boye and Arcand, 2013). Green growth is the pursuance of growth and development economically as well as prevention from degradation of environment, loss in biodiversity and unsustainable utilization of natural resources (Boye *et al.*, 2012). Inthe manufacturing of traditional packaging materials like paper, cardboard and plastic, a huge amount of energy is utilized therefore green packaging technology involves the methods that are environment sensitive. Commonly, the fossil fuel energy is utilized that produces a drastic amount of carbon dioxide and methane gases into the atmosphere and also the waste of that packaging material are dumped in landfills or water bodies cause environmental degradation (Qiang and Min, 2015).

For successful implementation of green technology, it is important that consumers should accept and trust that this eco-friendly packaging is safe and convenient. Consumer usually like to relate themselves with green products, if it fulfills their satisfaction level and as a consequence attains consumer loyalty (Yazdanifard and Mercy, 2011; Wei and Yazdanifard, 2013).

3.10 Green Packaging

The main aspect of green packaging should be balancing the relation between the food packaging and resource, environment, consumption of energy, the waste disposal and consumer's health and safety (Feng and Liu, 2011). Green packaging has emerged as an important concept to resolve the issues related to food packaging and its effect on environment that have emerged in last 50 years. In some developed countries, this advanced concept was referred as ecological packaging, non polluting packaging or environment friendly packaging. Since 1993, this concept was named green packaging in China. Green packaging is not clearly defined till now, but it is agreed to refer the packaging that can be recycled and degraded to conserve the resources and energy. The manufacturing of packaging material, packaging procedures to the waste disposal should be safe to environment and consumer health. Therefore, green packaging is a mean to conserve energy and resource to avoid waste and also recycle, reuse and decompose easily to fulfill the objective of protection of eco- friendly environment (Qiang and Min, 2015). Packaging normally complied with the 3R1D principles in most of the developed countries that represents reduce, reuse, recycle and degradable. Along with the sustained advancement, the latest demand of consumers is 4R1D in the green concept of packaging, in which a new R i.e. Refill is also added in the 3R1D principle (Feng and Liu, 2011; Qiang and Min, 2015).

The green package should be designed to prevent the wastage of natural resources and environmental pollution principally at every step of manufacturing of packaging material to the production, and even reuse, recycle and disposal. Green packaging does not only mean the packaging materials that may be easily decomposed but the fact should also be considered that manufacturing procedure for food packages should not contribute in the environment pollution and waste of resources and the packages could be recycled after use (Qiang and Min, 2015).

3.11 Biodegradable Packaging

Biodegradation refers to the process of decomposition of a polymeric material in the environment by the enzymatic action of the microorganisms within a very short time period. In this degradation process the chemical structure changes, mechanical and structural characteristics are lost, and as a result converted into various cleavage compounds like biomass, humic substances, carbon dioxide, water, minerals, etc. The biodegradation products and rate of biodegradation is affected by several factors which include temperature, humidity, presence of oxygen and pH (Zee, 2005; Jamshidian *et al.*, 2010). The biodegradable materials are cleavage materials that may be broken down, decompose in environment by the action of micro-organisms present in soil & water and light. These cleavage materials again enter into the ecological environment and hence return in the non-toxic ways to the nature (Qiang and Min, 2015).

Taking into consideration the harmful effect of petroleum derived plastics that cause environment pollution, the requirement for biodegradable packaging has increased in the food industries. Bio- polymers are bio-molecules that are commonly found in cellulose and proteins. These bio-polymers are manufactured from the sustainable resources like plant derived components such as plant cellulose, starch, sugar, plant based oil, etc. (Ambrose, 2020). Bio-polymers are manufactured from crude fats and oils, and other natural sources. Bio-polymers are categorized into four classes (Chandra and Rustgi, 1998; Clarinval and Halleux, 2005):

c) Bio-polymers that are derived from natural sources, such as carbohydrates (cellulose and starch); proteins (casein and gelatin); silk; aquatic unicellular organisms.

f) Bio-polymers that are manufactured by chemical process using biologically generated monomers (polylactic acid).

g) Bio-polymers that are synthesized by bacteria, commonly from genetically modified GM bacteria e.g. poly hydroxy alkanoates, poly hydroxy butyrate, hydroxyl valerate.

h) Bio-polymers that are manufactured from plant crude fat, such as PVA, modified polyesters, and modified polyolefin, that are temperature and light sensitive.

Based on historical development, biodegradable packaging materials are divided in three generations.

A. I - Generation

The first generation of material was utilized manufacturing of shopping bags that are generally produced from synthetic polymers as Low density polyethylene with starch (5 – 15 %) and pro-oxidizing and auto-oxidative additives. But such materials decompose into the molecules that are not biodegradable in nature (Chiellini, 2008).

B. II - Generation

The second generation of biomaterials contains a mixture of pre-gelatinized starch and LDPE (40 – 70 %) along with the hydrophilic co-polymer such as vinyl acetate,ethylene acrylic acid, and polyvinyl alcohol. The complete degradation of thesebiomaterials takes 2–3 years while only starch takes 40 days for degradation (Chiellini, 2008).

C. III - Generation

The third generation of the biomaterials is that consists of biomaterials completely. These are divided into three classes as per the origin and methods of production (Chiellini, 2008; Ivankovic *et al.*, 2017):

a) Polymers derived from biomass
b) Polymers prepared by chemical process and bio-monomers
c) Polymers that are derived from genetically modified bacteria or natural sources

Shanghai Chemical and Industrial Corporation has developed the biodegradable biaxially oriented polylactic acid film that was extracted from corn. This degradable film may be applied as food package with high glossiness, transparency, heat seal-ability, steady folding and good strength (Qiang and Min, 2015).

A biodegradable composite film was developed by mixing of chitosan and starch that consist good water vapour and mechanical properties. Chitin and chitosan having antimicrobial properties are found to be a good ingredient for production of biodegradable packaging for food products as it also help in prolonging the shelf-life of foods due to antimicrobial effect (Chiellini, 2008; Ivankovic *et al.*, 2017).

Biodegradable packaging is produced from the materials that are eco-friendly, therefore these can be easily recycled and require less energy in its production. These materials emit very less carbon and are nontoxic, hence very little or no effect on climate change. Along with many advantages of biodegradable packaging over plastics, these materials have some limitations also. As most of the biodegradable materials are produced from plant source, there will be more requirement of plant matter for their production in long term usage. Some of the single use biodegradable films are produced by mixing starch with small amounts of petroleum based polyester that is very hard to be decomposed by certain microorganisms. Such materials will require some special treatments for their decomposition. Another emerging issue is land filling as most of the bio-degradable materials require more than one year to decompose or compost (Pawar and Purwar, 2013).

Bio-degradable packaging may serve as a better substitute to plastics when utilized in combination with metal containers. This combination may provide the better solution for packaging of perishables. This concept will reduce the use of fossil fuel due to less dependence on plastics. However bio-degradable packaging hasn't attained its complete blossom yet to conserve the eco - system, therefore equitable consumption of such substitute is preferable (Ambrose, 2020).

3.12 Edible Packaging

Edible packaging is an innovative concept that will contribute to an advanced stage in the field of packaging of foods. This packaging concept will be helpful in reducing the amount of plastic packaging waste and also will be potential profitable to the stakeholders and the community (Deshwal and Panjagari, 2020). Edible film is referred to as fine, steady or intact sheet that ismanufactured from edible substances (Balasubramaniam *et al.*, 1997; Guilbert*et al.*, 1997). Edible packagings are excellent alternative to the traditional plastic packaging or biodegradable packaging. These environmental friendly films are manufactured from natural polymeric materials like protein (soy or milk proteins) and lipids (Eagle, 2004). Edible films may be put between foodmaterials, used as wrapper and also as a pocket for holding food products (Hernandez-Izquierdo and Krochta, 2008). Edible films have air barrier properties, provide protection physically and also serve as a good alternate for plastic particularly as wrapper packaging. Edible colours, flavours and antioxidants may also be added for various applications in the market (Bourtoom, 2009).

Edible films are completely eco-friendly and are not only safe to the consumer and environment but also helpful in reduction of packaging waste discarded from market and domestic waste. Its manufacturing practices are not too complicated. Its production is very convenient due to lower manufacturing cost and less processing time. These packaging are also advantageous for useras fine layer of film gets melted easily when it is mixed with saliva. Edible films may also be tailored as per requirements of the market. Edible films aretasteless, odorless, colorless, and may be manufactured in variety of colors, flavors and even sweetness. Besides, it also contributes nutrition to food products as it is composed of natural food components (Wei and Yazdanifard,2013).

Edible packaging are classified in two categories: one is in the form of edible paper food packaging prepared from starches, sugars and few condiments; second one is a type of edible scrap prepared by modification of edible fibers along with some food additives. Some edible packing films that are prepared by mixing starch and soy protein, that claims to maintain the water content and oxygen and also safeguard its freshness, taste and nutritive value. Some of the researches reveal that food packaging container can also be prepared with the bean curd residue that is rejected as a waste during processing of bean products. Complete exploitation of such waste as a resource may increase the quality of such materials and also keep away the problem of environment pollution (Qiang and Min, 2015).

The concept of edible packaging suggests a comprehensive path for consumption of various food products. Along with the green characteristics, edible package contribute robustness and flavor to food. Such green technologies have potential to manufacture good quality food products as it fetches a greater value than traditional ones (Wei and Yazdanifard, 2013).

3.13 Sustainable Food Packaging

World Commission on Environment and Development (WCED), 1987 have described the sustainable development that human society has the potential to make development sustainable by ensuring and conforming the current requirements without negotiating the ability of requirements for future. Sustainable development is defined as the degree of consumption and exercise that can be continued in the predictable future, as a result the systems that contribute products and services to consumers keep on endless (Robertson, 2007).

According to Lewis, 2008, the important principles of sustainability for food packaging are:

A. Effectiveness

Packaging should contribute worth to the society by containment and protection of products effectively as they run through supply chain and also support the system by proper information and responsible utilization.

B. Efficiency

The packaging should be properly fabricated to utilize energy and resources in efficient manner. Efficiency may be interpreted in the reference to the best exercise at each step of packaging cycle.

C. Cycling

Packing materials that are used, should continuously cycled with minimal degradation of material through the natural or industrial systems. Rate of recovery should be properly optimized for ensuring the energy and to save greenhouse gases.

D. Safe

Components of packaging should be pollution free and have no toxic effect. This includes raw materials, finished material, ink, colouring components and various additives that cause no risk to ecosystems or consumers. Precautionary principles should be applied in case of any doubt.

The preferences and demands of consumers are growing for clean, healthy, resource efficiency, less wastage and increased sustainable products and services. The interest of public is increasing in regard to environment pollution, wastage, safety & health and the requirement for this approach. Therefore, sustainability is assumed as a better formula for this advanced approach, however, the concept is not too understandable. Therefore, there is a great burden over industries to manufacture more healthy and sustainable produce and better services (Robertson, 2007).

3.14 Current and Future Challenges

In the field of food packaging, innovations are focused on the user friendly and practical aspects along with affability and aesthetic quality for consumers. Some green innovations claim for sustainability on the basis of bio - resources or by biodegradability without assessing their potential benefit on environment. Most of the green innovations do not fulfill the eco-friendly criteria as per expectations with reference to quantity and quality of renewable sources or degradability (Guillard *et al.*, 2018). To improve the sustainability of food package, there is a need of well equipped R&D and innovative packaging technologies (Davis and Song, 2006).

A lot of research has been done in the field of development of green and eco-friendly solutions for food packaging. However, packaging industries are facing the difficulties in specific technical issues regarding these green packaging materials, that currently hinder large market uptake. These technical issues are commonly related to variation in raw materials, processing conditions, diffusivity, permeability and composting, and also lack of tools to tailor packaging as per the needs of the consumers (Guillard *et al.*, 2018).

3.15 Conclusion

Food packaging is not packaging or a product only, however it is a systematic approach and therefore packaging is always considered as a integral part of its content. In context with conservation of natural resources and environment protection, the best solution for reduction of food packing wastes is the utilization of green materials for packaging and implementation of green and eco-friendly techniques in food packaging. A lot more research needs to be done to utilize more natural materials for food packaging so that the issue regarding the disposal of the plastic packaging may be reduced.

3.16 Reference

Ambrose DCP. 2020. Biodegradable Packaging – An Eco-Friendly Approach. Curr. Agri. Res., 8, 1, 04-06, .

Balasubramaniam VM, Chinnan MS, Mallikarjunan P, Philips RD. 1997. The effect of edible film on oil uptake and moisture retention of deep-fat fried poultry product. Journal of Food Process Engineering, 20, 1, 17–29.

Barlow C and Morgan D. 2013. Polymer film packaging for food: An environmental assessment. Resources, Conservation And Recycling, 78, 74-80.

Bourtoom, T. 2009. Review article edible protein films: properties enhancement. International Food Research Journal, 16, 1-9.

Boye JI and Arcand Y. 2013. Current Trends in Green Technologies in Food Production and Processing, Food Eng. Rev., 5, 1–17.

Boye JI, Maltais A, Bittner S, Arcand Y. 2012. Greening of research and development. In: Green technologies in food production and processing, Boye JI, Arcand Y (eds), Springer, New York.

Bugusu B. 2007. Food Packaging and Its Environmental Impact, Food Technology Magazine, Volume 61, No. 4.

Chandra R and Rustgi R. 1998. Biodegradable polymers. Prog Polym Sci., 23,1273–335.

Chiellini E. 2008.Enviromentally compatible food packaging. Woodhead publishing limited, Cambridge England, 8–10.

Clarinval AM and Halleux J. 2005. Classification of biodegradable polymers. In: Biodegradable polymers for industrial applications. Smith R, editor. 1st ed. Boca Raton, FL, USA: CRC Press. p 3–31.

Davis G and Song JH. 2006. Biodegradable packaging based on raw materials from crops and their impact on waste management. Ind. Crops Prod., 23,147–61

Deshwal G and Panjagari N. 2020. Review on metal packaging: materials, forms, food applications, safety and recyclability, J. Food Sci. Technol., 57, 2377–2392.

Dey A, Dhumal CV, Sengupta P et al. 2020. Challenges and possible solutions to mitigate the problems of single-use plastics used for packaging food items: a review. J Food Sci Technol., 2020, https://doi.org/10.1007/s13197-020-04885-6

Eagle S. 2004. Edible wraps. Food in Canada, 64, 1, 24.

Feng M and Liu M. 2011. On the importance of green package used in food industry. J. Taiyuan Sci., 2, 41- 43.

Guilbert S, Cuq B, Gontard N. 1997. Recent innovations in edible and/or biodegradable packaging materials. Food Additives and Contaminants, 14, 6, 741–751.

Guillard V, Gaucel S, Fornaciari C, Angellier-Coussy H, Buche P, Gontard N. 2018. The Next Generation of Sustainable Food Packaging to Preserve Our Environment in a Circular Economy Context. Front. Nutr. 5,121.

Hernandez-Izquierdo VM and Krochta JM. 2008. Thermoplastic processing of proteins for film formation - A review, Journal of Food Science, 73, 2, 30–39.

Ivankovic A, Zeljko K, Talic S, Bevanda AM, Lasic M. 2017. Biodegradable packaging in the food industry, Journal of Food Safety and Food Quality, 68, 2, 23–52.

Jamshidian M, Tehrany EA, Muhammad Imran, Jacquot M, Desobry S. Poly-Lactic Acid: 2010. Production, Applications, Nanocomposites, and Release Studies, Comprehensive Reviews in Food Science and Food Safety, 9, 552-571.

John PJ. A 2013. Handbook on Food Packaging, Astral International.

Kooijman JM. 2000. Environmental Impact of Packaging: Performance in the Household, INCPEN, London.

Lewis H. 2008. Eco-design of food packaging materials. In: Environmentally - Compatible Food Packaging, Chiellini E (ed.), Cambridge, England: Woodhead Publishing Ltd, 238–262.

Mesias F, Eldesouky A, Pulido A. 2015. The Role of Packaging and Presentation Format in Consumers‘ Preferences for Food: An Application of Projective Techniques. Journal of Sensory Studies. 30.

Pawar PA and Purwar AH. 2013. Biodegradable Polymers in Food Packaging, American Journal of Engineering Research, 2, 5, 151-164.

Pongracz E. 2007. The Environmental Impacts of Packaging, In: Environmentally Conscious Materials and Chemicals Processing, Kutz M, John Wiley & Sons, 237-277.

Qiang W and Min Z. 2015. Research on the Food Green Packaging Under the Sustainable Development, Adv. J. Food Sci. Technol., 7, 12, 910-913.

Robertson GL. 2007. Sustainable food packaging, Handbook of waste management and co-product recovery, Woodhead Publishing Series in Food Science, Technology and Nutrition, Pages 649-662.

Wei LT and Yazdanifard R. 2013. Edible Food Packaging as an Eco-friendly Technology using Green Marketing Strategy, Global Journal of Commerce and Management Perspective G.J.C.M.P., 2, 6, 8- 11.

Williams H, Wikström F, Otterbring T, Löfgren M, Gustafsson A. 2012. Reasons for household food waste with special attention to packaging. Journal of Cleaner Production, 24, 141-148.

Yazdanifard R and Mercy IR. 2011. The impact of green marketing on customer satisfaction and environmental safety. 2011 International Conference on Computer Communication and Management, 5, 637-641.

Zee MV. 2005. Biodegradability of polymers–Mechanisms and evaluation methods. In: Handbook of biodegradable polymer. Bastioli C, editor. 1st ed. Shropshire, U.K.: Rapra Technology Limited. 1–22.

4

Entrepreneurial Opportunities Through Value Addition of Jackfruit (*Artocarpus Heterophyllus*)

Sapna Arora* **and** ***Anjali Shankhwar***

ABSTRACT

The ancient fruit "Jackfruit" is one of the most remunerative and important underutilized native fruits of India. It is extremely versatile sweet tasting fruit and vegetable rich in nutritional value. Modern technologies are evolved for the development of value-added products from jackfruit is studied in the present study. To prevent from post-harvest losses the wine production from jackfruit juice which is a very good substrate for fermentationcan be done at larger scale. Popularity of jackfruit from the limited regions can be explored over the country. There is a good scope and possibility of marketing the value added products in the country. Value addition is carried out by minimal processing by changing its form to a valuable product having more shelf life. The developed products enhance good income opportunities for farmers and also entrepreneurship opportunities for small and medium enterprises for the popularization of jackfruit to a wider region.

Keyword: Jack Fruit, Value Addition, entrepreneurial opportunities.

4.1 Introduction

The word "Jackfruit" derived from Malayalam language "Chakka" and comes from Portuguese word jaca. Jack (Artocarpusheterophyllus) is a tropical fruit species belongs to family Moraceae.Usually, it is grown in sea shoreline, heavy rainfall and humid areas of the world.

It is the much loved fruit of many, due to its sweetness. The jackfruit plant is also called as "poor man's food" as it is economical and availability in abundant during the season. The jackfruittree is widely cultivated in humid

*Corresponding Author

areas of India and other Asian tropical countries like Bangladesh, Nepal, Sri Lanka, Vietnam, Thailand, Malaysia, Indonesia and the Philippines (Devet al., 2014). Vietnam is the world leader in jack fruit based value added products. Jackfruit is alsofound across African countries e.g., in Cameroon, Uganda, Tanzania, and Mauritius, as well asthroughout Brazil and Caribbean nations such as Jamaica. Still, India is considered as the native place of jack fruit. Jackfruit tree is evergreen, long lived having a life span of more than 60 years with a height of 10 to 15 meter (Dattatraya 2012).

The ripe jackfruit is having sweet taste and has an exotic flavor. The whole Jackfruit consist of mainly three parts i.e. bulb (30-32%), seeds (18%) and the rind (5-55%) (Srivastava et al., 2017).

Usually, people used to eat fresh, used in ice crème preparation or put into salad the ripe jackfruit. After minimal processing or dehydrated; sometimes ripe jack fruit can be sold as dried pulp and further, various value added products can be prepared like chutney, jam, jelly, pastes, Bar, nectar, pickle etc. People consume boiled jack fruit seeds or roasted one due to its richness in starch. Some part of jackfruit is considered as inedible because of underdeveloped perianths which usually cover the bulbs part of the jackfruit.

As a commercial crop, the jackfruit is not popular due to more variations in the quality even if it has numerous advantages. Further, people have some beliefs that too much consumption of jackfruit flakes also results into some digestive problems in future. Moreover, it's difficult to separate out fruit bulbs from rind.

Table 1: State wise value of output of Jackfruit in India

State-wise Value of Output from Jackfruit in India {(At Constant 2011-12 Prices) (2011-2012 to 2017-2018)} (Rs. in Lakh)							
States/UTs	2011-2012	2012-2013	2013-2014	2014-2015	2015-2016	2016-2017	2017-2018
Arunachal Pradesh	133	133	132	132	143	108	70
Assam	19445	22638	19444	19759	19917	19976	20118
Bihar	1153	1329	1776	0	17032	17032	0
Jharkhand	1007	1754	1757	1777	10241	10879	12019
Karnataka	13459	13498	12516	12879	12382	11912	12335
Kerala	34468	38016	37256	37129	36115	35608	39156
Manipur	60	36	0	0	0	0	0
Mizoram	167	165	167	166	166	166	166
Nagaland	57	68	259	259	208	263	264
Odisha	0	0	0	13337	13339	13352	13352
Puducherry	0	0	0	0	0	5	15
Tamil Nadu	24950	8293	9121	23973	20981	60395	33061
Tripura	21761	23850	26752	26559	25814	13290	11995
Uttar Pradesh	900	968	864	1025	1310	1344	1400
West Bengal	50145	50536	50798	51189	51389	51934	54617
India	**167704**	**161284**	**160840**	**188184**	**209038**	**236264**	**198570**

Source: Indiastat.com, Ministry of Statistics and Programme Implementation, Govt. of India.

4.2 Nutritional Properties

Jackfruit can act as a source of complete nutrition due to nutrition availability in it. When looking for mixture of nutrition, it is equivalent to Avocado and Olive for dietary needs. It is having exact nutrient equivalents of mother's milk. Jackfruit is abundant in some vitamins like B and C and minerals such as potassium, calcium, iron. It is also rich in proteins and high level of carbohydrates in India. Jackfruit contains a chemical "Jacalin" which is useful for the prevention of colon cancer, AIDS etc. This fruit also rich in proteins and can be relished as a nutritious nut.

When it comes to the composition of jack fruit (100 gram edible portion), water content is on higher side in young fruit (76.2 – 85.2g) as well as ripe fruit (72.0-94.0 g). The other nutrients are also on higher side like Calcium (30.0 – 73.2 mg for young fruit; 20.0-37.0 for Ripe fruit), Phosphorous (20.0 – 57.2 mg for young fruit; 38.0- 41.0 for Ripe fruit), Potassium (287 – 323 mg for young fruit; 191- 407 mg for Ripe fruit). Total sugar and Magnesium is absent in young fruit whereas, both are present in 20.6 g and 27.0 mg respectively. Further, jackfruit provides different level of energy in case of young fruit (50-210 KJ) and ripe fruit (88-410 KJ). Moreover, the availability of Vitamin A is also into different range in case of young (30 g) and ripe fruit (175 -540 gm).

4.3 Health Benefits

Jackfruit is full of protein, minerals, Vitamin C and other nutritions however, people are still unaware about its health benefits. The jackfruit protects against cancer due to presence of the phytonutrients like lignanats, liso flavones and saponin. Due to richness in fibre, and presence of unique stick form, jackfruit works as a colon cleanser and remove toxin from digestive tract. (Swami and Kalse, 2019).The presence of Vitamin C in jackfruit helps in strengthens the immune system. Further, Jackfruit also contains anti-ulcer properties and the presence of high fiber(3.6 g/100 g) prevents constipation which helps in healthy digestion. The jackfruit also boosts energy level due to presence of simple sugars like fructose and sucrose which provides immediate energy. The availability of raisin or latex on Jackfruit trunk and fruit can be helpful for cleaning solution for dental purposes and also useful for dental filling material in future. (Swami and Kalse, 2019).

4.4 Value Addition and Entrepreneurial Opportunities of Jackfruit

Jackfruit has great potential for value addition by changing or transforming a product from its original state to a more valuable state. Value addition results in –

- Minimization of post-harvest loss
- Non seasonal availability
- Increment of the economic value
- Consumer appeal of an agricultural commodity

Chips

Raw or unripe jackfruits are used for chips preparation. Fruit is sliced into suitable sizes and blanched into hot water at 95°C for 5 minutes and dried at 60°C for an hour followed by 70°C for next 6 hours (Srivastava *et al.*, 2017). Finally the slices are fried at 160°C in edible oil and seasoned with salt and packed. Vietnam is the world leader in vacuum fry jackfruit chips.

Pickle

Unripe fruit are sliced into small pieces (especially bulbs and seeds), mixed with salt, oil and spices and kept for aging.

Dehydrated Jackfruit

The unripe bulbs are blanched and dehydrated for the further use throughout the year.

Ready to Cook (RTC) Jackfruit

Fruit is properly washed, peeled and sliced into suitable sizes. The pieces preserved by minimal processing to prepare ready to cook jackfruit products.

Brined Jackfruit

Mature fruits are washed properly with clean water, peeled and sliced into suitable pieces. Pieces are mixed with 8 percent salt, 1.25 percent acetic acid, 0.1 percent KMS and 91.65 percent water solution (Srivastava *et al.*, 2017). Mixture is poured in air tight plastic container and stored in cool and dry place.

Jackfruit Candy

Half ripe fruit is washed properly and sliced into 0.5x0.5 cm size and blanched in hot water at 95 °C for 4 minutes(Srivastava*et al.*, 2017). Pieces are dipped in 2 percent calium lactate and 0.1 percent KMS for 2 hoursand drained finally. The pieces are further dipped into 25, 25, 45, 50, 60 and 70 Brix into sugar solutions. The slices are washed with clean water to remove adhering syrup and dried at 70 °C in cabinet dryer up to 10 percent moisture content. The final product is packed in polypropylene pouch and stored at room temperature.

Nectar

Pulp made by passing bulbs in pulping machine and mixed with 10 percent hot water are used to make nectar.

Jackfruit Leathers

Bulbs from fully ripe fruit washed properly and blended with 10 – 15 percent sugar and boiled for 5 to 7 minutes for juice extraction. The mixture is boiled for 3 to 5 minutes after addition of KMS at rate of 0.1 g/kg. The mixture is concentrated with steam jacketed pan and dried for 20 hours at 60°C to amoisture content of 20 percent (Srivastava *et al.*, 2017). Finally cooled mixture sliced into desirable sizes and stored after packaging.

Jackfruit Jam

Ripe fruit bulbs are blended and boiled for 5 to 7 minutes to extract juice. 700 gm. of sugar and 10 gm. pectin is added to 1 kg fruit pulp and cooked till TSS reaches to 64° Brix and finally 0.25 percent citric acid is added. After determination of the end point by flake test jam is poured in sterilized bottles and stored at room temperature.

Jackfruit Jelly

Ripe fruit washed properly and rind is separated and sliced into small pieces. For 1 kg rind 1.5 lit water and 2g citric acid is added and mixture is boiled for 35 minutes and the juice is extracted from the rind.700g sugar and 200 mg citric acid is added to the extracted juice and cooked till the TSS reaches to 65 °Brix followed by addition of citric acid (Srivastava *et al.*, 2017). The final mixture is poured into sterilized bottle and stored at room temperature.

Ready to Eat (RTE) Jackfruit Products

Ripe jackfruit bulbs after minimal processing are preserved and can be converted into ready to eat convenience food product have a limited shelf-life and can be stored and transported under refrigerated conditions.

Jackfruit Wine

The three main steps in jackfruit wine making are

Preparation of Seed Culture

- Fresh jackfruit washed thoroughly, seeds are removed.
- The fruit is cut into fine pieces (229.5 g of the pulp, 558 ml of water and 112.0 g of sugaris added) and blended in a blender to obtain jackfruit juice.

- The juice filtered and transferred to a conical flask and autoclaved in an autoclave at 121°C for 15 minutes to inhibit the growth of unwanted micro-organisms.
- In the extracted juice 0.3 g yeast extract is inoculated to facilitate the growth of yeast cells (*S. cerevisiae*).
- The extracted juice is incubated at 30°C on rotary shaker (150 rpm) overnight.

Preparation of Sample/Must

- Jackfruit pulp, water and jaggery is added and mixed well (blended) to obtain must/ sample.
- 250 mL of the must transferred into each pre-sterilized bottle for fermentation.
- The pH of the must adjusted to 4, 5, and 6 for each temperature and inoculum concentration.
- The bottles are autoclaved again at 121°C for 15 minutes and inoculated culture of concentration of 5%, 10% and 15% to each bottle at pH of 4, 5, and 6 respectively.
- Samples were taken out on 7th, 14th and 21st day for the testing parameters like pH and Inoculums concentration.

Table 2: Value Addition of Jackfruit

Part of Fruit	Value Addition
Unripe fruit	Pickle, Chips, Papad, Brined Jackfruit, RTCJackfruit, Dehydrated Jack-fruit, Culinary preparations,Cutlets, Biryani.
Half ripe fruit	Candy, Preserve
Fully ripe fruit	Jam, Leather, Rind Jelly, Squash, Nectar, Canned Bulbs, RTE (Ready-to-eat) Bulbs, RTS(Ready-to-serve) Drinks, Chutney , Toffee, Wine, Halwa, Kheer, GulabJamun, Icecream, Custard, Cake, Freeze dried pulp.
Seeds	Seed powder, Starch flour, Culinary preparation,Pakoda, Kheer

In India, jackfruit is grown mainly in homestead farms and has multiple uses like food for human beings, feed for animals. The increment in the production at high level can give opportunities for the farmers and industries can be established to produce different jackfruit products. Scientific and economic reasons are the important emerging factors can help government to promote and commercialize jackfruit products. Nowadays jackfruit is widely used for medicinal purpose as it contains secondary metabolites having biological activities and different researches are also are in progress to explore its value in different sectors. It is required to develop some strategic plans based on the results of different researches for conservation of this KalpaVriksha.

Value addition to jackfruit is still not growing in India due to less available products for consumers and inferiority complex of the people. Jackfruit seed powder can be as a substitute of refined wheat flour which is economical and also seeds can be used as a source of protein to address malnutrition problem.

By utilizing different government schemes marketing network jackfruit-based products can be manufactured at different plants in the country and can be export to different countries. The new plants utilize technology is developed by the CFTRI and ICAR which helps in manufacturing a range of products like candies, jam, syrup and nectar for the domestic and export markets.Establishment of a mechanism for thc procurement of jackfruit from farmers through the Vegetable and Fruit Promotion Council Kerala (VFPCK) can also show good results in decreasing post- harvest losses.Further, through value addition of the jackfruit, the earnings of the stakeholder can be boost up to 10,000 crore (Vinayak, 2020).

4.5 Conclusion

The objectives of the present study were to study the value addition of jackfruit by different processing techniques.Jackfruit can play a significant role in the livelihood security through enhancing household income and employment generation. Beingrich in nutritional, medicinal and processing qualities,there is a great scope of establishing large scale jackfruit processing industries in the country. The significant production of this fruit ensures availability of the fruit throughout the year and the emerging technologies for processing and value addition of jackfruit need to be promoted and commercialized to avoid the wastage of this fruit.

4.6 References

Arora T, Parle A. Jackfruit - A health Boon. Int J Ayurveda Pharma. May-June, 7(3), 59-64, 2016.

Baliga, M.S.,Shivashankara, A R, Haniadka R, Dsouza J, Bhat, H P. Phytochemistry, nutritional and pharmacological properties of Artocarpusheterophyllus Lam (jackfruit): A review, Food Research International, 44, 1800–1811, 2011. doi: 10.1111/j.1541-4337.2012.00210.x.

Gunasena, H P M, Ariyadasa, K P, Wikramasinghe A, Herath H M W, Wikramasinghe, P, Rajasearuna S B, Manual of Jack Cultivation in Sri Lanka. Forestry Information Service. Forest Department, 48, 1996.

https://www.indiastat.com/agriculture-data/2/stats.aspx

Kushala G, Sreenivas KN, VasudevaNaik K and Ranjani SN. Product development acceptability and cost effectiveness of jackfruit jam blended with avocado and kokum, Journal of Pharmacognosy and Phytochemistry, 1043-1045, 2017.

Molla, M M, Nasrin T A A, Islam M N and Bhuyan M A J. Preparation and packaging of jackfruit chips. Int. J. Sustain. Crop Prod. October, 3(6), 41-47, 2008.

Mondal C and Remme R N. Product Development from Jackfruit (Artocarpusheterophyllus) and Analysis of Nutritional Quality of the Processed Products. IOSR Journal of Agriculture and Veterinary Science, Jul. – Aug, 4 (1), 76-84, 2013.

Ranasinghe R A S N, Maduwanthi S D T, Marapana R A U J. Nutritional and Health Benefits of Jackfruit (Artocarpusheterophyllus Lam.): A Review; International Journal of Food Science, 2019. https://doi.org/10.1155/2019/4327183.

Saxena T, Saxena A, Raju P S. Development of value added products from jackfruit for small and medium enterprises. Indian Food Pack. March-April, 67, 95-104, 2013.

Sharma N and Bhutia S P. Process Optimization for Fermentation of Wine from Jackfruit (Artocarpusheterophyllus Lam.), J Food Process Technol, 4:2, 2013. DOI: 10.4172/2157-7110.1000204.

Sidhu A S. Jackfruit Improvement in the Asia-Pacific Region– A Status Report. Asia-Pacific Association of Agricultural Research Institutions. 2012. https://www.apaari.org/wp-content/uploads/downloads/2012/10/Jackfruit-A-Success-Story_31-8-2012.pdf accessed on 14.12.2020.

Soepadmo E, Artocarpusheterophyllus lam. In: Verheij, E.W.M., Coronel, R.E. (Eds.), Plant Resources of Southeast Asia No. 2: Edible Fruits and Nuts. PROSEA, Wageningen, Netherlands, pp. 86–91, 1992.

Srikant, B S; Thakor, N J; Haldankar, P. M. and Kalse, S. B. Jackfruit and its many functional components as related to human health: A review. Comprehensive Reviews in Food Science and Food Safety, 11, 565-576, 2012.

Srivastava, A; Bishnoi, S.K. and. Sarkar, P K. Advances in value addition in jackfruit for food and livelihood security of rural communities of India. The Asian Journal of Horticulture. June, 12(1), 160-164, 2017.

Sudheer, KP, Saranya, S. Jackfruit: A new hope for entrepreneurship and value addition, Ind Food Ind Mag, July-August, 36(4), 26-31, 2017.

Swami, S. B. and Kalse, S B. Jackfruit (Artocarpusheterophyllus): Biodiversity, Nutritional Contents, and Health.. J.-M. Mérillon, K. G. Ramawat (eds.), Bioactive Molecules in Food, Reference Series in Phytochemistry, https://doi.org/10.1007/978-3-319-78030-6_87. pp 2237-2258, 2019.

Vinayak A J,Expert stresses need for value addition to jackfruit, The Business Line, Chennai Edition, 3rd September 2020.

Waghmare R, Memon, N, Gat Y, Gandhi S, Kumar, V and Panghal, A. Jackfruit seed: an accompaniment to functional foods. Braz. J. Food Technol., 22, e2018207, 2019. https://doi.org/10.1590/1981-6723.20718.

5

Ultrasound Assisted Drying and Its Impact on Bioactive Compounds of Fruits and Vegetables

Swati Mitharwal, Sachin Kumar, Neerja Usha Kujur
Komal Chauhan, Prabhat Kumar Nema*

ABSTRACT

Fruits and vegetables are rich source of bioactive compounds such as polyphenols, antioxidants and flavonoids. Drying is one of the common methods for preservation of food and agricultural commodities. However, drying may negatively affect the product quality attributes. This article is a capsule overview of current state of art ultrasound (US) assisted drying on technical aspects and bioactivity of fruits and vegetables. US assisted drying is an economical and sustainable technique compared to other existing conventional techniques used for food processing and preservation. US as a pretreatment before drying of agricultural-products enhances drying efficiency, inactivate enzymes and improve overall quality of dried products. Current developments and available solutions in drying have shown that US assisted drying may be used to restrict the loss of crucial bioactive components occurring naturally in fruits and vegetables. US when combined with hot air drying helps in achieving adequate drying rate at lower temperatures and improves overall dehydration process. This technique has found no practical application commercially due to various difficulties in achieving high-intensity US in air. One practical benefit of using US assisted drying is the minimization of the drying time along with reduction in total energy consumption.

Keywords: Ultrasound assisted drying, fruits and vegetables, poly-phenolic compounds, antioxidant activity, plant pigments.

5.1 Introduction

Foods like fruits, vegetables and aquatic products are highly prone to degradation because of its high water activity and heat sensitivity. Food

***Corresponding Author**

dehydration is the most commonly used method to enhance the shelf-life of fruits and vegetables. However, the technology employed for food dehydration should be efficient, economic and capable of protecting the nutrients, color, appearance, texture and flavor. But in many cases it is found that this unit operation greatly hampers the product quality (Zhang et al., 2017). Drying is an energy intensive and time consuming process so there is a need for new drying techniques. In this section we are focusing on ultrasonic-assisted drying.

US may increase the drying process of fruits and vegetables by increasing the mass transfer intensification from the surface layer of food without significant increase in temperature. Ultrasound energy is generated by mechanical sound waves that are above the threshold of human hearing (>16 kHz). US is an efficient pre-processing step which intensify and accelerate the drying process. The US has an ability to preserve the bioactive components occurring naturally in fruits and vegetables (Siucińska & Konopacka, 2014). US has various industrial applications including the food technology and processing industry such as minimization of processing, maximize quality and ensure food safety as well. It is used to improve food preservation, mass transfer, assistance of thermal treatments, modify texture and food analysis. Based on the frequency range employed ultrasound can be divided into two types, low and high energy. US techniques are simple, energy saving and economical, thus became a promising novel technology in food sector. US applications includes microbial inactivation, drying, extraction, freezing as well as processing of cereal products, vegetables and fruits, meat product, food proteins and food enzymes.

US may be utilized to test various quality parameters including internal defects in fruits, texture, and composition. The ultrasound is generally used in combination along with other thermal treatments to attain adequate drying rates at lower processing temperatures (Gallego-Juárez et al., 2007). So these treatments provide a synergistic effect and help to reduce the undesirable effects of heating (Karel & Lund, 2003).

There is a need to develop and design effective power ultrasonic systems that could be applied to large scale operations. Different applications of ultrasound cannot be generalized, and thus it should be custom designed for each specific application. The selection of appropriate ultrasound system depends on physicochemical and functional properties of food (Awad et al., 2012).

5.2 Principle of Ultrasound Technology

The medium of propagation of US waves is either through bulk of the material or through its surface and the speed of travel depends on the nature of the wave or the material of propagation. There are various frequency ranges for ultrasound. Until recently, high frequency waves ranging between 100 kHz to 1 MHz and power less than 1 W cm-2 are most commonly used for non-destructive analysis and the quality assessment of food material. Low-intensity US provides information regarding the physicochemical characteristics of fruits and vegetables such as carbohydrate, degree of maturity, acidity etc. It is also used in genetic improvements for live stocks, check the quality of meat, fish, and poultry.

The low frequency (16-100 kHz) and high-power US waves in the range of 10 to 1000 W cm-2 is employed to change the physicochemical characteristics of food material. Low frequency high power US is also known as power ultrasound. Power US brings about physical, chemical and mechanical changes through cavitation process, which help in the extraction of various bioactive components from food, control microstructure, drying, freezing, thawing, fat sonocrystallization (modification of fat texture), improve functional properties of food protein, drying, defoaming, emulsification, food surface sanitization and enhancement of shelf life of food.

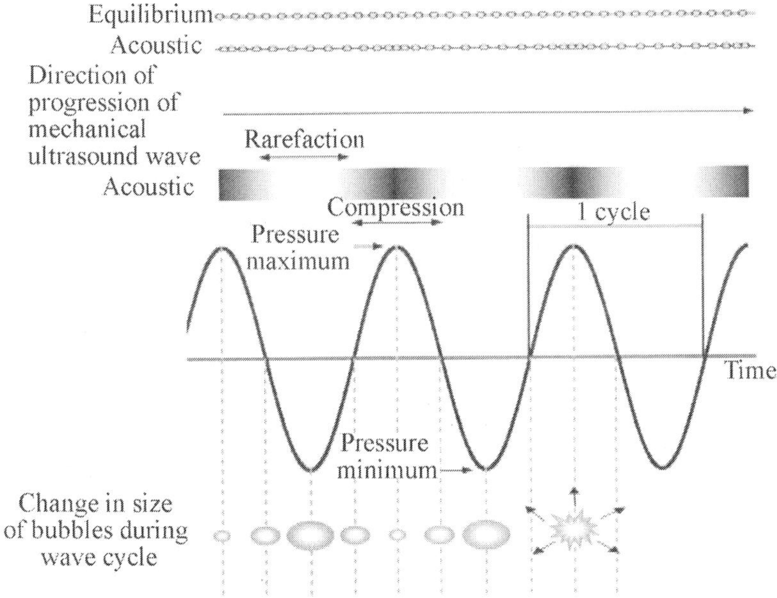

Fig. 1: Ultrasonic cavitation (adapted from Soria 2010)

US acoustic energy waves propagates through a series of compression and rarefaction (Fig 1). This propagation generates cavitation bubbles that are formed from gas nuclei existing within the fluid. These cavitation bubbles spread all over the liquid medium and grow until the critical size is reached beyond which they become unstable and further collapse. This explosion of bubbles induces rise in temperature (5000K) and pressure (1000 atm) leading to generation of very high shear energy and turbulence with in cavitation zone. The extent of cavitation depends on viscosity of medium, surface tension, pressure, heat, vapor pressure, dissolved gas, energy, and intensity.

The explosive implode of cavitation bubbles generates strong micro stream currents, shear stresses and high velocity gradients which alters the media characteristics. The localized pressure and temperature increase but temperature is generally lower than 70°C. This may lead to breakdown of water molecules and generation of highly reactive free radicals which reacts and brings out changes in other molecules. This US mechanism is responsible for altering physicochemical properties of food products (Soria & Villamiel, 2010).

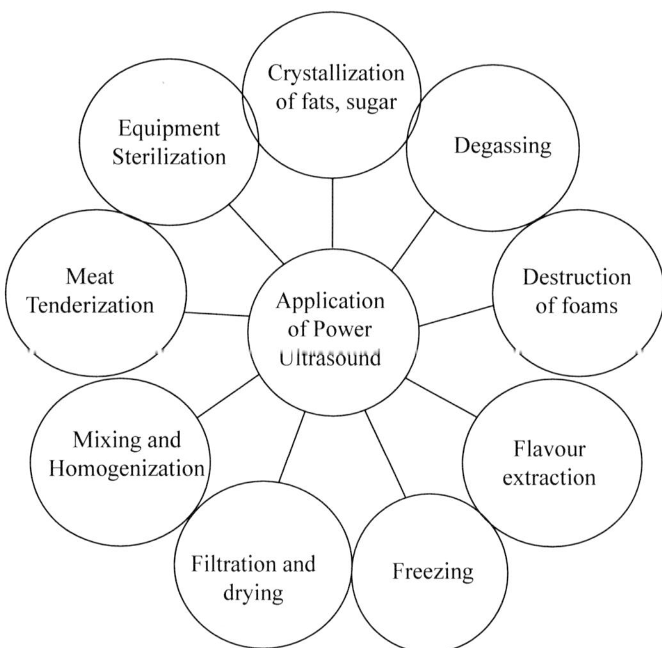

Fig. 2: Applications of power ultrasound in food processing

5.3 Ultrasound as Drying Pretreatment

The fruits and vegetables when treated with power ultrasound brings about physicochemical changes in their tissue structure that leads to increase in water diffusivity and reduction in drying time. The travel of ultrasonic waves through a medium lead to series of compression and rarefaction waves that is similar to sponge effect. It generates microscopic channels which facilitates removal of moisture. These microscopic channels are used by water molecules to move towards the surface of fruits and vegetables. The main principle behind US acoustic wave energy is to minimize the diffusion boundary layer and augmentation of convective mass transfer within the sample (Fernandes *et al.*, 2011).

The ultrasound pretreatment is performed under ambient temperature conditions as high temperature inhibits the desirable effects. The principle of cavitation process work against the naturally occurring osmotic pressure gradient that expels water out of the fruit. High moisture fruits and vegetables releases greater amount of soluble solids to the liquid medium than the low moisture fruits and vegetables. US influences soluble solids diffusion from the tissue structure of fruits and vegetables. The mass transfer between the fruit and corresponding medium depends on void size of microscopic channels. The dimension of microscopic channels greatly depends on cultivar of fruits and vegetables. The fruits treated with ultrasonic pretreatment have shown a reduction in air drying time as compared to fresh fruits without any pretreatment (Delgado, Zheng, & Sun, 2009).

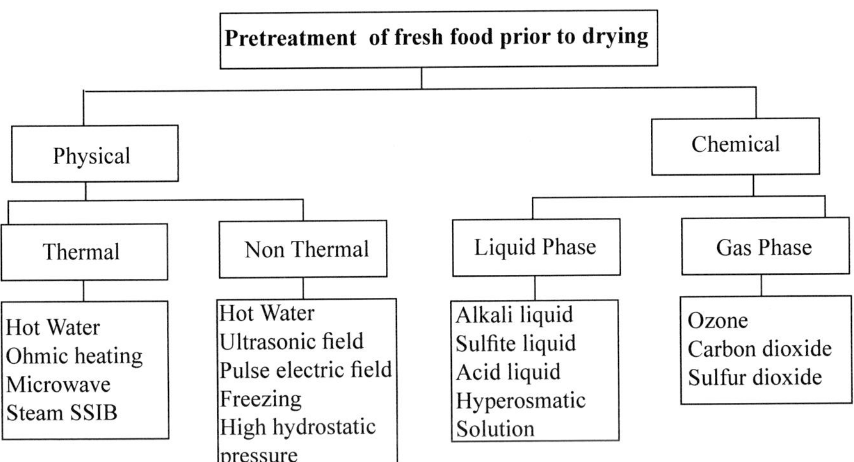

Fig. 3: Classification of drying pretreatments used in food processing

5.4 Intensification of Mass Transfer

US is a prominent pretreatment used prior to drying which plays a pivotal role in mass transfer, enhanced water loss and solute gain. US waves reduces the medium temperature from 70°C to 40°C without any change in diffusion parameters. The intensity of US pretreatment affects water and solute transport. In hypertonic (30° Brix) sucrose solution, US application shortened the drying time by 8-40% (Nowacka *et al.*, 2012).

In a study conducted on ultrasound assisted drying of melon found that US modifies the cell structure without bringing any significant change in cell thereby enhancing the mass transfer phenomena. The presence of microscopic voids on cell structure causes an increase in water diffusivity as tissue offer low hinderance to water diffusion. US pretreatment is used to reduce the drying time, while drying fruits or vegetable high in moisture content. US pretreatment in association with high sucrose concentrations improve the appearance of strawberry and make its color bright and vivid (Fernandes & Rodrigues, 2007).

5.5 Effect of Ultrasound on Drying Process

Ultrasound waves may be directly connected with the drying equipment. For smooth transfer of acoustic energy from transducer to air and then to dried material, US generating equipment is generally connected with several devices. The samples are placed on the vibrating elements which is in direct contact with the transducer. US acoustic energy positively influences the drying process. The US energy provides various mechanical effects on both the material being dried and the gas-solid interface. The application of US may accelerate water removal process without increasing the temperature during drying providing alternate technique for processing heat-sensitive food materials (Garcia-Perez *et al.*, 2012).

The US waves have been shown to minimize internal and external resistance to mass transfer during drying process. The effectiveness of US not only depends on wave frequency, pressure, and sound intensity, but also on experimental conditions. The transfer of UV energy is greatly affected by raw material properties and air velocity. There is an inverse relationship between air velocity and acoustic field. Furthermore, the transfer of moisture depends on the pore size of the raw material. García-Pérez *et al.* (2007) found that in the case of highly porous products such as lemon peel, US energy is able to impact the internal mass transfer processes for the entire air velocity range under study. On the other hand, for less porous products like carrots, the effectiveness of US is limited to lower air velocity only.

During the application of US, the drying rate depends on the mass loading capacity of the drying chamber. The changes associated with application of US affects the surface wax distribution of fruits or vegetable tissue thereby altering the moisture migration and convective mass transfer (Ortuño *et al.*, 2010).

The energy consumption of any US-assisted drying could be minimized by utilizing intermittent application of ultrasound. US processing causes cell injuries leading to structural deformations at tissue level associated with water removal. It was found that US application does not alters the color, rehydration characteristics, vitamin C and other heat sensitive food components. Also, foods with higher porosity are more susceptible to US as compared to less porous foods. US is a promising technique for reducing drying time and temperature for processing heat-sensitive foods thereby maximizing the quality of dried products.

5.6 Effect of Ultrasound Assisted Drying on Bioactive Compounds

(i) Chlorophyll

Chlorophyll is a valuable bioactive compound that is extracted from marine micro algal biomass. It has antioxidant and anti-mutagenic properties. Chlorophyll may be divided into two parts, chlorophyll a and chlorophyll b. The chlorophyll is made up of porphyry in macro cycle that is composed of four pyrrole rings which consists of 4C and 1 N atom. It is abundant in nature and due to its active role in photosynthesis, it is crucial for the survival of plants and animals (Hosikian, Halim, & Danquah, 2010). Spray drying technique has been found to be an effective method for preservation of chlorophyll using whey protein isolates as a wall material (Zhang *et al.*, 2019).

Roueita, Hojjati, and Noshad (2020) investigated effect of application of osmotic solutions (sugar, grape and mulberry syrups) on chlorophyll content of kiwi fruits treated with ultrasound assisted drying (27 KHz, 65°C) for 20-40 mins. The results revealed that chlorophyll content decreased with increase in ultrasound treatment time. This decrease was ascribed to leaching and breakdown of chlorophyll pigment. Further, chlorophyll loss was significantly lower in kiwifruit slices treated with osmotic solutions (0.14-0.56 mg/kg) as compared to samples treated with distilled water (0.11-0.34 mg/kg). The loss was lowest in sample treated with grape syrup. The reduction in chlorophyll's loss was due to enhanced viscosity of aqueous phase which affected the chlorophyll's reduction diffusion. Nowacka et al. (2017) studied the effect on chlorophyll content of ultrasound assisted (35 kHz, 10-30 mins) osmotic dried kiwi fruit. The results revealed that chlorophyll content of kiwifruit depends on ultrasound exposure time and was highest for samples US treated for 20 mins while it was 85.4 mg/kg for fresh kiwifruit. This increase in chlorophyll content was ascribed to release and better extraction of pigments and bioactive compounds upon US application. Jin, Zhang, and Shi (2019) evaluated chlorophyll content of ultrasound (1200 W, 15 mins) pretreated radio frequency assisted hot air-dried (RFHAD) bitter melon. The chlorophyll content was better retained in RFHAD dried sample (1.54 mg/g) as compared to hot air-dried samples (1.22 mg/g).

(ii) Carotenoids

Carotenoids are secondary metabolites which belongs to the class of isoprenoid pigments and are synthesized by plants, bacteria, and fungi. Over six hundred carotenoids have been identified and are classified as xanthophylls (astaxanthin, lutein and zeaxanthin) and carotenes (β-carotene and lycopene). Carotenoids are highly unstable and susceptible to degradation during food processing such as drying, cooking and storage (Saini & Keum, 2018).

The impact of ultrasound assisted air drying (45-60°C, 75 W, 21 kHz) on carotenoid content of cherry tomatoes was studied by Fernandes et al. (2015). Ultrasound treatment at milder conditions resulted in better preservation of carotenoids and lycopene. Nowacka, and Wedzik (2017) investigated the impact of ultrasound pretreatment on carotenoid content of carrots dried using convective drying.

The study outcome showed that ultrasound treatment (21 kHz, 20 mins) increased the carotenoid content by 12.5%. Further, samples treated at higher ultrasound frequency (35 kHz) resulted in 41.1- 49.9% increase in carotenoid content as compared to raw carrots (7.66 mg/100g). This increase in carotenoid content was due to disintegration of cellular structure which further facilitated its extraction.

(iii) Vitamins

Vitamins are organic compounds that plays vital role in human metabolism and normal body functioning. They can be classified in two groups, water soluble vitamins (Vitamin C and B complex) and fat-soluble vitamins (Vitamin A, D, E and K). Vitamins are sensitive towards temperature, air, light, and moisture which affects its stability during processing (Mamede et al., 2011). Impact of ultrasound assisted drying on the profile of vitamins have been studied extensively by several researchers. Gamboa-Santos et al. (2014) studied the effect of ultrasound treatment (40-70°C, 0-60 W) on vitamin C retention of strawberries. Vitamin C retention was higher in samples treated at lower temperatures (upto 50°C). Furthermore, the treatment at higher temperature combined with ultrasound reduced the processing time but had detrimental effect on vitamin C retention. However, this retention was higher as compared to samples processed by convective drying alone. Effect of ultrasound assisted (40 kHz, 0-300 mins) osmotic dehydration on vitamin C of radish cylinders was carried out by Xu et al. (2014). The ascorbic acid content of all the processed samples was significantly lower as compared to fresh radish sample (29.36 mg/100g). However, the retention was better in ultrasound treated samples rather than control. The decrease in vitamin C content during ultrasound processing may be due to reaction with hydroxyl radicals or thermal degradation. Zhang et al. (2016) quantified changes in vitamin C content of hot air dried button mushroom slices treated with ultrasound (40 kHz, 3-10 mins). Ultrasound application reduced the drying time by 9.5% as compared to control. Ascorbic acid was found in range of 1.09-1.12 mg/100g DW as compared to untreated sample (1.10 mg/100g DW). The impact of ultrasound assisted air drying (45-60°C, 75 W, 21 kHz) on vitamin content of cherry tomatoes was investigated (Fernandes et al., 2016). The study showed that ultrasonication positively influenced the vitamin A and E content and the retention was higher in samples treated at lower temperature and air velocity (45°C and 1 m/s). The better retention of lipid soluble vitamins on ultrasound treatment was due to increased drying rate. Further, US treatment resulted in release of bounded B vitamins thereby improving its bioavailability.

(iv) Phenolic Compounds

Polyphenols are the group of secondary metabolites that are abundantly distributed in plant kingdom and are found to exhibit several bioactivities such as alleviation of blood glucose level inhibiting starch digestibility in vivo (Sun & Miao, 2020). Drying significantly alters the cell wall composition and cellulose microfibril structures thereby reducing the binding capacities of polyphenols (Liu *et al.*, 2019).

Several researchers have investigated the impact of ultrasound treatment on total phenolic content of fruits and vegetables dried using various techniques (Table 1). Pirce *et al.* (2020) examined the effect of ultrasound (0-130 kHz) and osmotic pretreatment (0-70° Brix) on total polyphenol content (TPC) of murtilla skin fruit dried by convective drying. The TPC of ultrasound treated samples were significantly higher than non-sonicated samples while it was

715.3 mg/100g for fresh fruit. This rise in TPC content might be due to release of polyphenolic compounds and addition of hydroxyl group in polyphenolic aromatic ring during ultrasound treatment. Kroehnke et al. (2018) quantified effect of ultrasound assisted (26 kHz, 75-200 W) convective drying on TPC of carrots. The TPC of fresh sample (1102 mg GAE/ kg) was significantly higher than dried samples. Further, ultrasound treatment significantly increased TPC as compared to control sample which may be due to reduction in drying time thereby minimizing polyphenolic compounds decomposition. High profile liquid chromatography (HPLC) analysis of carrot extracts revealed chlorogenic acid and isocoumarin as major polyphenolic compounds followed by ferulic acid, hydroxycinnamic acid derivatives and caffeic acid. Amami *et al.* (2017) worked on ultrasound assisted (40 kHz, 10-30 mins, 20-40°C) osmotic dehydration pretreatment on TPC of dried strawberry. The results suggested that TPC loss was lesser (38%) in samples treated with ultrasound at lower temperature as compared to those processed at higher temperature of 60°C (51.94%). The rise in TPC content may be because of generation of pores in plant tissue as a result of ultrasound processing which gave better extractability of bioactive compounds.

(v) Flavonoids

Flavonoids are the largest classes of secondary metabolites found in different parts of plants such as seeds, leaves, roots, stem etc. They are found to possess several health benefits such as antioxidant, anti-inflammatory, anticancer activity and have protective effects against coronary heart and kidney diseases (Górniak, Bartoszewski, & Króliczewski, 2019). The rapid oven drying pretreatment at 75°C produced the greatest loss of flavonoids. Zhang et al. (2019) found initial pretreatment in the shade as effective method to preserve flavonoids followed by oven drying at 75°C. Various previous research studies have been conducted to study the application of ultrasound waves as pretreatment to conventional and non-conventional drying of fruits and vegetables (Table 2). Ren *et al.* (2018) studied the impact of ultrasound pretreatment (20 kHz, 1-5 mins) on total flavonoid content (TFC) of onion slices dried by hot air drying. The treatment at high ultrasound amplitude (61 μm) resulted in better retention of TFC as compared to control sample.

Sonication induces cavitation and higher shear forces thereby increasing mass transfer and resulting in improved extractability of flavonoid compounds. Liu *et al.* (2019) investigated effect of contact ultrasound application on TFC content of pear slices dried using fair- infrared radiation (FIR). It was found that ultrasound pretreatment positively influenced the TFC content of samples from 62.11-82.1 mg TE/100g to 70.30-89.22 mg TE/100g at FIR power (100-340 W). This trend was ascribed to reduced drying time with ultrasound application. Rashid *et al.* (2019) worked on ultrasound pretreatment on flavonoid content of sweet potatoes dried by infrared drying. The results revealed that samples treated at medium range ultrasound frequency of 40 kHz and drying temperature of 60°C had significantly higher TFC content (8.5 mg Ru/g) as compared to control sample.

(vi) Antioxidant activity

Antioxidants are the chemical compounds that have ability to scavenge or neutralize free radicals formed in the body. The uncontrolled activity of free radicals can cause oxidative stresses, which may break down various macromolecules (DNA, protein, lipids) leading to diabetes, carcinogenesis, cardiovascular diseases, and accelerated ageing (Khan *et al.*, 2019). Chen *et al.* (2019) studied the effect of different drying methods (hot air, freeze, vacuum, and spray drying) on antioxidant activity of bamboo shoot residue derived polysaccharides (CPS). Freeze dried CPS was found to exhibit better antioxidant activity which may be ascribed to its lower glucose content, lower molecular weight and higher uronic acid concentration. The impact of ultrasound as a pretreatment for drying of fruits and vegetables on antioxidant profiles was reported by several previous studies. Rashid et al. (2019) studied the effect of ultrasound pretreatments on antioxidant activity of sweet potatoes. The results showed that dried samples had significantly higher antioxidant activity as compared to fresh sweet potatoes sample which may be related to increase in concentration of polyphenolic compounds and enhanced TPC, TFC. Further, samples dried at higher temperature (80˚C) and ultrasound frequency of 60 kHz showed stronger 2,2-diphenyl-1-picrylhydrazyl (DPPH) antioxidant activity (9.51 µmol TE/100 g) and Ferric Reducing Antioxidant Power (FRAP) activity (2.50 mmol Fe/g). Baeghbai et al. (2018) quantified antioxidant activity of apple slices dried by ultrasound assisted conductive hydro-drying. Drying process decreased antioxidant activity significantly as compared to fresh sample and the decrease was lower in samples pretreated with ultrasound (6.37 mg Trolox/g) as compared to untreated samples (4.83 mg Trolox/g DM). The higher retention of antioxidants may due to effect of ultrasonication on sample texture resulting in better extraction of antioxidant compounds. Cárcel *et al.* (2019) studied the impact of ultrasound treatment on antioxidant properties of red pepper. The results indicated the detrimental effect of drying on antioxidant activity of red pepper (89-97% loss). Further, antioxidant degradation for sample treated at higher temperature (70˚C) and ultrasound waves was lower as compared to those treated at lower temperatures (30˚C). This might be due to reduced drying time which compensated for negative effect associated with acoustic energy.

Table 1: Effect of US assisted drying on total phenolic content (TPC)

Sample	Drying method	Drying Conditions	TPC	References
Carrots slices	US Drying	DTe: 20-60^{0}C DT: 75-120 mins USF: 20 kHz	Dried: 1.111-1.366 mg GAE/g DM	Soria et al., 2010
Strawberry (semi spherical halves)	Pretreatment: US assisted OD Drying: Convective drying	USTe: 20, 30, 40^{0}C UST: 10-30 mins USF: 40 kHz DT: 40, 50, 60^{0}C AV: 1.0 m/s	Fresh: 27.77 mg GAE/g DM Dried: 13.34-17.31 mg GAE/g DM	Amami et al., 2017
Onion slices	Pretreatment : US assisted HAD	USTe: 70^{0}C UST: 1-5 mins USF: 20kHz DTe: 60^{0}C DT: 8 h AV: 0.3 m/s	Fresh: 7.76 mg GAE/g DM Dried: 6.50-8.58 Mg GAE/g DM	Ren, Perussello, Zhang, Kerry, and Tiwari, 2018
Apple slices	US CHyD	DTe: 25^{0}C USF: 25 kHz	Fresh: 6.37 mg GAE/g DM Dried: 6.51 mg GAE/g DM	Baeghbali, Niakousari, Ngadi, and Eskandari, 2019
Bitter Melon slices	US assisted HAD	Pretreatment: USP: 1200 W UST: 15 min USTe: 25^{0}C HAD Conditions DTe: 50^{0}C AV: 1.0 m/s	Control (HAD): 32.64 mg GAE/g Dried: 34.98 mg GAE/g	Jin, Zhang, and Shi,2019
Pear slices	US assisted FIRD	USF: 28 kHz USP: 0-60 W FIR P: 100, 220, 340W AV: 1.5 m/s DTe: 30^{0}C	Dried: 195-13-268.31 mg GAE/g DM	Liu et al., 2019
Red pepper (square pieces)	US VD	DTe: 45-75^{0}C AV: 1.3 m/s	Dried: 1.404- 2.194 mg GAE/g	Tekin, and Baslar, 2018

		P: 15 mbar	DM	
Broccoli (small pieces)	US assisted AD	USF: 20 kHz USI: 125.2, 180.1 W/dm^2 DT: 70^{0}C	Fresh: 11.09 mg GAE/g DM Dried: 7.95-7.97 mg GAE/g DM	Cao et al., 2019
Litchi pulp	US assisted HAD	USP: 750 W DT: 10 mins DTe: 65^{0}C AV: 1.8 m/s	Dried: ~500 mg/kg DM	Cao et al., 2019
Pear slices	US assisted HAD	AV: 1 m/s USF: 28 kHz DTe: 35, 45, 55^{0}C USP: 0, 24, 48W	Fresh: 572.35 mg GAE/100g DM Dried 408.88-467.02 mg GAE/100g	Liu, Zeng, Wang, Sun, and Xi, 2019
Cranberries chunks	US assisted CD	USF: 21kHz USP; 180 W DTe: 23^{0}C 61.5% Sucrose solution	Fresh: 4349 mg GAE/100g Dried: 926-1157 mg GAE/100g	Wiktor et al., 2019
Sweet potatoes slices	US assisted IFD	USF: 20,40,60 kHz AV: 0.2 w/cm^2 DTe: 60-80^{0}C	Fresh: 1.91 mg GAE/g Dried: 6.18-15.70 mg GAE/g	Rashid et al., 2019
Honeyberry fruits	US assisted VD	USF: 37 kHz UTe: 40^{0}C	Fresh: 6.209 g GAE/100g Dried: 2.216 g GAE/100g	Žlabur et al., 2019
Murtilla fruit	US assisted CD	USF: 0-130 kHz DTe: 70^{0}C AV: 2 m/s DT: 5-6 h	Fresh: 715.3 mg/100g DM Dried ~ 400 mg/100g DM	Pirce et al., 2020

*abbreviation: US, ultrasound; OD, osmotic dehydration; HAD, hot air drying; CHyD, conductive hydro-drying; FIRD, far-infrared radiation drying; VD, vacuum drying; CD, convective drying; IFD, infrared drying; AD, air drying; GAE,gallic acid equivalent; DT: drying time; DTe, drying temperature; USTe, ultrasound treatment temperature; UST, ultrasound treatment time; AV, Air velocity; USF, ultrasound frequency; USI, ultrasound intensity; USP, ultrasound power.

Table 2: Effect of US assisted drying on total flavonoid content (TFC)

Sample	Drying method	Drying Conditions	TFC	References
Onions slices	Pretreatment : US assisted HAD	USTe: 70°C UST: 1-5 mins USF: 20kHz DTe: 60°C DT: 8 h AV: 0.3 m/s	Fresh: 3.34 mg QE/g DM Dried: 3.35-4.34 mg QE/g DM	Ren, Perussello, Zhang, Kerry, and Tiwari, 2018
Apple slices	US CHyD	DTe: 25°C USF: 25 kHz	Fresh: 1.86 US HD: 1.75 mg CE/g DM	Baeghbali, Niakousari, Ngadi, and Hadi, 2019
Pear slices	US assisted FIR Dryer	USF: 28 kHz USP: 0-60 W FIR P: 100, 220, 340W AV: 1.5 m/s DTe: 30°C	Dried: 70.30-89.22mg TE/g DM	Liu *et al.*, 2019
Pear slices	US assisted HAD	AV: 1 m/s USF: 28 kHz DTe: 35, 45, 55°C USP: 0, 24, 48W	Dried: 130.46-211.41 mg/100g	Liu, Zeng, Wang, Sun, and Xi, 2019
Sweet potatoes slices	US assisted IFD	USF: 20,40,60 kHz AV: 0.2 w/cm^2 DTe: 60-800C	Fresh: 1.27 mg Ru/g Dried: 2.74-8.55 mg Ru/g DM	Rashid *et al.*, 2019
Mushroom slices	US assisted CD	AV: 1 m/s USF: 22 kHz USP: 50 W DTe: 5-15°C	Dried: ~8-10 mg GAE/g DM	Vallespir, Crescenzo, Rodríguez, Marra, and Simal, 2019
Honeyberry fruits	US assisted VD	USF: 37 kHz UTe: 40°C	Fresh: 2.825 g GAE/100g Dried: 0.746 g GAE/100g	Žlabur *et al.*, 2019

*abbreviation: US, ultrasound; OD, osmotic dehydration; HAD, hot air drying; CHyD, conductive hydro-drying; FIRD, far-infrared radiation drying; VD, vacuum drying; CD, convective drying; IFD, infrared drying; AD, air drying; GAE,gallic acid equivalent; DT: drying time; DTe, drying temperature; USTe, ultrasound treatment temperature; UST, ultrasound treatment time; AV, Air velocity; USF, ultrasound frequency; USI, ultrasound intensity; USP, ultrasound power.

5.7 Conclusion

This article reveals a comprehensive and selective overview carried out on US assisted drying for dehydration of fruits and vegetables. This article gives importance to the mechanisms behind the cavitation process involved in the dehydration process. The probes used are economical, portable, and easy to modify as per different requirements. In general, fruits and vegetables are dried by utilizing the techniques suitable for heat sensitive materials to retain their color, texture, appearance, and nutrition quality. US assisted drying techniques are economically profitable due to low maintenance and energy costs.US pretreatment releases bound phenolic compounds and flavonoid thereby enhancing the antioxidant activity. Further, US assisted drying resulted in better retention of pigments and vitamins as compared to conventionally dried fruits and vegetables. Finally, ultrasound assisted drying is not a standard technique, so it requires further research for large scale industry applications. Furthermore, US may increase temperature of the dried materials when it is applied at a higher power for an extended period of time.

5.8 References

Amami E, Khezami W, Mezrigui S, Badwaik LS, Bejar AK, Perez CT, Kechaou N. 2017. Effect of ultrasound-assisted osmotic dehydration pretreatment on the convective drying of strawberry. Ultrason Sonochem, 36, 286-300.

Awad TS, Moharram HA, Shaltout OE, Asker D, Youssef MM. 2012. Applications of ultrasound in analysis, processing and quality control of food: A review. Food Res. Int., 48(2), 410-427.

Baeghbali V, Niakousari M, Ngadi MO, HadiEskandari M. 2019. Combined ultrasound and infrared assisted conductive hydro-drying of apple slices. Dry. Technol, 37(14), 1793-1805.

Cao Y, Tao Y, Zhu X, Han Y, Li D, Liu C, Liao X, Show PL. 2019. Effect of microwave and air-borne ultrasound-assisted air drying on drying kinetics and phytochemical properties of broccoli floret. Dry. Technol, 1-6.

Cárcel JA, Castillo D, Simal S, Mulet A. 2019. Influence of temperature and ultrasound on drying kinetics and antioxidant properties of red pepper. Dry. Technol, 37(4), 486-493.

Chen G, Li C, Wang S, Mei X, Zhang H, Kan J. 2019. Characterization of physicochemical properties and antioxidant activity of polysaccharides from shoot residues of bamboo (Chimonobambusa quadrangularis): Effect of drying procedures. Food Chem, 292,281-293.

Delgado AE, Zheng L, Sun DW. 2009. Influence of ultrasound on freezing rate of immersion-frozen apples. Food Bioproc Tech, 2(3), 263-270.

Fernandes FA, Rodrigues S, Cárcel JA, García-Pérez JV. 2015.Ultrasound-assisted air-drying of apple (Malus domestica L.) and its effects on the vitamin of the dried product. Food Bioproc Tech, 8(7), 1503-1511.

Fernandes FA, Rodrigues S, García-Pérez JV, Cárcel JA. 2016. Effects of ultrasound-assisted air-drying on vitamins and carotenoids of cherry tomatoes. Dry. Technol, 34(8), 986-96.

Fernandes FA, Rodrigues S, Law CL, Mujumdar AS. 2011. Drying of exotic tropical fruits: a comprehensive review. Food Bioproc Tech, 4(2), 163-185.

Fernandes FAN, Rodrigues S. 2007. Ultrasound as pre-treatment for drying of fruits: Dehydration of banana. J. Food Eng, 82(2), 261-267.

Gallego-Juárez JA, Riera E, De la Fuente Blanco S, Rodríguez-Corral G, Acosta-Aparicio VM, Blanco A. 2007. Application of high-power ultrasound for dehydration of vegetables: processes and devices. Dry. Technol., 25(11), 1893-1901.

Gamboa-Santos J, Montilla A, Soria AC, Cárcel JA, García-Pérez JV, Villamiel M. 2014. Impact of power ultrasound on chemical and physicochemical quality indicators of strawberries dried by convection. Food Chem, 161, 40-46.

García-Pérez JV, Cárcel JA, Benedito J, Mulet A. 2007. Power ultrasound mass transfer enhancement in food drying. Food Bioprod Process, 85(3), 247-254.

Garcia-Perez JV, Carcel JA, Riera E, Rosselló C, Mulet A. 2012. Intensification of low-temperature drying by using ultrasound. Dry. Technol, 30(11-12), 1199-1208.

Górniak I, Bartoszewski R, Króliczewski J. 2019. Comprehensive review of antimicrobial activities of plant flavonoids. Phytochem Rev, 18(1), 241-272.

Hosikian, A., Lim, S., Halim, R., & Danquah, M. K. 2010.Chlorophyll extraction from microalgae: a review on the process engineering aspects. Int. J. Chem. Eng.

Jelena S and Juan LS. 2007. Influence of osmotic concentration, continuous high frequency ultrasound and dehydration on antioxidants, colour and chemical properties of rabbiteye blueberries. Food Chem,101 (3), 898-906.

Jin W, Zhang M, Shi W. 2019. Evaluation of ultrasound pretreatment and drying methods on selected quality attributes of bitter melon (*Momordica charantia* L.) Dry. Technol, 37(3), 387-396.

Karel, M., & Lund, D. B. 2003. Physical principles of food preservation: revised and expanded. CRC Press.

Khan IT, Nadeem M, Imran M, Ullah R, Ajmal M, Jaspal MH. 2019. Antioxidant properties of Milk and dairy products: a comprehensive review of the current knowledge. Lipids Health Dis, 18(1), 41.

Kroehnke J, SzadzińskaJ, Stasiak M, Radziejewska-Kubzdela E, Biegańska-Marecik R, Musielak G. 2018. Ultrasound- and Microwave-Assisted Convective Drying of Carrots – Process Kinetics and Product's Quality Analysis, UltrasonSonochem.

Liu D, Lopez-Sanchez P, Martinez-Sanz M, Gilbert EP, Gidley MJ. 2019. Adsorption isotherm studies on the interaction between polyphenols and apple cell walls: Effects of variety, heating and drying. Food Chem, 282, 58-66.

Liu Y, Sun C, Lei Y, Yu H, Xi H, Duan X. 2019. Contact ultrasound strengthened far-infrared radiation drying on pear slices: Effects on drying characteristics, microstructure, and quality attributes. Dry. Technol, 37(6), 745-758.

Liu Y, Zeng Y, Wang Q, Sun C, Xi H. 2019. Drying characteristics, microstructure, glass transition temperature, and quality of ultrasound strengthened hot air drying on pear slices. J. Food Process. Preserv, 43(3), 13899.

Mamede A C, Tavares S D, Abrantes A M, Trindade J, Maia J M, and Botelho M F. 2011. The role of vitamins in cancer: a review. NutrCancer, 63(4), 479-494.

Nowacka M, Wiktor A, Śledź M, Jurek N, Witrowa-Rajchert D. 2012. Drying of ultrasound pretreated apple and its selected physical properties. J. Food Eng, 113(3), 427-433.

Nowacka M, Wedzik M. 2016. Effect of ultrasound treatment on microstructure, colour and carotenoid content in fresh and dried carrot tissue. Appl Acoust, 103:163-171.

Óscar R, Valeria E, Carmen R, Antoni F, Juan AC, Susana S. 2018. Application of power ultrasound on the convective drying of fruits and vegetables: effects on quality. J. Sci. Food Agric, 98(5), 1660-1673.

Ortuño C, Pérez-Munuera I, Puig A, Riera E, Garcia-Perez JV. 2010. Influence of power ultrasound application on mass transport and microstructure of orange peel during hot air drying. Phys. Procedia, 3(1), 153-159.

Pirce F, Vieira TM, Augusto-Obara TR, Alencar SM, Romero F, Scheuermann E. 2020. Effects of convective drying assisted by ultrasound and osmotic solution on polyphenol, antioxidant, and microstructure of murtilla (UgnimolinaeTurcz) fruit. J. Food Sci. Technol.

Rashid MT, Ma H, Jatoi MA, Wali A, El Mesery HS, Ali Z, Sarpong F. 2019. Effect of infrared drying with multifrequency ultrasound pretreatments on the stability of phytochemical properties, antioxidant potential, and textural quality of dried sweet potatoes. J. Food Biochem, 43(4), 12809.

Ren F, Perussello CA, Zhang Z, Kerry JP, Tiwari BK. 2018. Impact of ultrasound and blanching on functional properties of hot-air dried and freeze dried onions. LWT, 87, 102-111.

Roueita G, Hojjati M, Noshad M. 2020. Study of Physicochemical Properties of Dried Kiwifruits Using the Natural Hypertonic Solution in Ultrasound-assisted Osmotic Dehydration as Pretreatment. Int. J Fruit Sci, 16, 1-7.

Saini RK, Keum YS. 2018. Carotenoid extraction methods: a review of recent developments. Food Chem, 240, 90-103.

Siucińska K, Konopacka D. 2014. Application of ultrasound to modify and improve dried fruit and vegetable tissue: A review. Dry. Technol, 32(11), 1360-1368.

Soria AC, Corzo-Martinez M, Montilla A, Riera E, Gamboa-Santos J, Villamiel M. 2010. Chemical and physicochemical quality parameters in carrots dehydrated by power ultrasound. J. Agric. Food Chem, 58(13), 7715-7722.

Soria AC, Villamiel M. 2010. Effect of ultrasound on the technological properties and bioactivity of food: a review. Trends Food Sci Technol, 21(7), 323-331.

Sun L, Miao M. 2020. Dietary polyphenols modulate starch digestion and glycaemic level: A review. Crit Rev Food Sci Nutr, 60(4), 541-555.

Tekin ZH, Baslar M. 2018. The effect of ultrasound-assisted vacuum drying on the drying rate and quality of red peppers. J. Therm Anal Calorim, 132(2), 1131-1143.

Vallespir F, Crescenzo L, Rodríguez Ó, Marra F, Simal S. 2019. Intensification of low-temperature drying of mushroom by means of power ultrasound: effects on drying kinetics and quality parameters. Food Bioproc Tech, 12(5), 839-851.

Wiktor A, Nowacka M, Anuszewska A, Rybak K, Dadan M, Witrowa Rajchert D. 2019. Drying kinetics and quality of dehydrated cranberries pretreated by traditional and innovative techniques. J. Food Sci, 84(7), 1820-1828.

Xu B, Zhang M, Bhandari B, Cheng X. 2014. Influence of ultrasound-assisted osmotic dehydration and freezing on the water state, cell structure, and quality of radish (Raphanus sativus L.) cylinders. Dry. Technol, 32(15), 1803-1811.

Zhang Z, Liu Z, Liu C, Li D, Jiang N, Liu C. 2016. Effects of ultrasound pretreatment on drying kinetics and quality parameters of button mushroom slices. Dry. Technol, 34(15), 1791-1800.

Zhang ZH, Peng H, Ma H, Zeng XA. 2019. Effect of inlet air drying temperatures on the physicochemical properties and antioxidant activity of whey protein isolate-kale leaves chlorophyll (WPI-CH) microcapsules. J. Food Eng, 245, 149-156.

Zhang X, Wang X, Wang M, Cao J, Xiao J, Wang Q. 2019. Effects of different pretreatments on flavonoids and antioxidant activity of Dryopteris erythrosora leave. PLoS One, 14(1), e0200174.

Zhang M, Chen H, Mujumdar AS, Tang J, Miao S, Wang Y. 2017. Recent developments in high-quality drying of vegetables, fruits, and aquatic products. Crit Rev Food Sci Nutr, 57(6), 1239-1255.

Žlabur JŠ, Colnar D, Voća S, Lorenzo JM, Munekata PE, Barba FJ, Dobričević N, Galić A, Dujmić F, Pliestić S, Brnčić M. 2019. Effect of ultrasound pre-treatment and drying method on specialized metabolites of honeyberry fruits (*Lonicera caerulea* var. *kamtschatica*). Ultrason Sonochem, 56, 327-372

6

Green Approach in Food Processing and Production Through Nanotechnology

Aryasree Sukumar* and *K.A. Athmaselvi*

ABSTRACT

Green technologies are required in the field of food processing and production in order to achieve food security and to reduce adverse environmental impact. Application of nano technology in food processing and production can advance the food safety and quality, increase the bioavailability of nutrients and thereby improving human health. Nanotechnology can be applied to achieve green perspective like development of biodegradable packaging materials thereby reducing adverse environmental effect. Use of naturally occurring nanoparticles as additives can reduce the use of chemicals in food products. This chapter deals with the potential of nanotechnology to achieve green aspects in food processing and production.Main focusses are on achieving food security and environmental sustainability. Chapter also describes about the principles of nanotechnology and methods to achieve green aspectin food sector with the application of nanotechnology. Application of nanotechnology in food processing and production along with safety, regulations and accompanying laws are discussed in this chapter.

Keywords: Nano Technology, Green Technology, Food Processing, Nano Materials, Food Nanoparticles, Nanoemulsion, Nano packaging, Nanosensors

6.1 Introduction

Food sector has important role in economic growth of a country and its sustainability(Srivastava *et al.*, 2018). Food is essential for sustaining healthy life for human kind. Energy required for sustaining life is derived from metabolism of food. Due to increase in population, there is increasing demand for production of safe and healthy foods. To meet global requirement

*Corresponding Author

productivity of crops, have to be increased from 60% to 100 % (Giraldo *et al.*, 2019). To get higher yield in this adverse environmental condition, new technologies have to be implemented. Implementing proper monitoring system for food hazards like pathogens, chemical residues and additives can help in achieving this goal (Omanovic-Miklicanina and Maksimovic., 2016). Through nanotechnology technological innovation in food sector can overcome these challenges. Different fields in food sector where nanotechnology can implement positive impacts are food safety, food security, new food product/ingredient development, increasing the absorption of nutrients, targeted delivery of functional food, development of nutraceuticals, packaging innovations and technologies to increase shelf life of food products (Ranjan *et al.*, 2014). Targeted delivery of active components through encapsulation technique is a new achievement of nanotechnology in food sector. Quality and quantity of crop and food products can be increased through nanotechnology. Food additives made of nanomaterials are found to increase shelf life of the food product, also used to enhance texture, color, flavor and nutrient composition of food product. Nanomaterial based additives can be used for the detection of spoilage, microbial growth and contamination in the product. Nanoparticles like zinc oxide, silver and titanium dioxide can be used as antimicrobial agent in food products. Nanotechnology has a progressive positive impact on diary and agricultural industry. Another important sector is food packaging, through the use nanomaterial infused packaging material, safety of the food product can be ensured. It can also detect spoilage and control other quality parameters like color and flavor. Shelf-life extension through release of preservatives is another mile stone in food sector. Nanoparticle incorporated biosensors help in monitoring storage conditions of grain storage system and also food products.

6.2 Food Security and Environmental Sustainability

Nanotechnology is a modern technique that is recently being applied to food processing sector and is having a positive effect on food industry. Application of nanotechnology can improve the physical parameters of food products like taste, texture and also increase the shelf life of food products. Nanotechnology applied methods are used as delivery technique for bioactive ingredients. It is also applicable in the field of food packaging and pathogen detection. Overall application of nanotechnology in food processing results in development of environmentally safe methods for improving the nutritional value of food (Grumezescu and Oprea *et al.*, 2017).

6.3 Nanotechnology

Nanotechnology is a technique that deals with macromolecules in the size range of 1-100nm and gained importance in food sector in recent years.

Nanotechnology can be applied to create material with novel properties. New product created using nanotechnology possess external or internal dimension the range of 1 to 100nm (Sharon *et al.*, 2010). Thus, nanotechnology possess a potential to modernize food system. Properties like bioavailability, safety, nutritional value and efficiency can be upgraded using nanoscale level of foods. Major application of nanotechnology in food and agricultural sector are advancing food processing and food security, increase ability of plants to absorb nutrients, aid in nutrient delivery techniques, improving flavor and nutrition of foods, detection of pathogen, increasing the food functionality, environmental safety, reducing cost of storage and distribution. Benefits of nanotechnology are in the development of new material, it aids in micro and nanoscale processing, new product development, help in designing new methods and instrumentation technique in food processing (Ghouri *et al.*, 2020).Important field of application of nanotechnology in food sector are as follows:

- Biosensors for the detection of specific food borne pathogens
- Nano film or coating formation and preservation of food product
- Nanoparticle which acts as antimicrobial to kill antibiotic resistant microbes thus preserving food
- Nanosensors based food packaging system to detect spoilage and extension of shelf life of food product
- Encapsulation of active material using nanoemulsion to aid in targeted delivery, core material stabilization and controlling rheological property of product (Rahmati *et al.*, 2020).

A. Areas of Application

(i) Food Processing

Nanotechnology plays important role in food processing now a days. Different fields of application in food processing are enzyme support system, delivery technique to enhance nutrient absorption, enzyme immobilization, processing technique to increase nutrient bioavailability,and nanoencapsulation techniques. Enzymes are applied in food processing to increasing nutritional value of product, enhancing flavor and also for health benefits. Nanomaterial can help as a support system for enzyme delivery thus increasing its shelf life, activity and reducing cost of production than that of traditional macro scale support system(Yu *et al.*, 2005). Forhydrolyzing olive oil, Nano silicon dioxide particles can be used which improves its stability, reusability and adaptability(Bai *et al.*, 2006).For immobilization of enzymes magnetic nanoparticles, Nano porous media, Nano fibers and carbon nanotubes are being used (Hwang 2013). Immobilized lipase enzymes on nanofibers and nanoparticles are used for hydrolysis of soya bean oil (Verma *et al.*, 2013). Nanoparticles are added to food during processing to improve nutrient absorption. These particles are designed in such a way that they break and release nutrients only upon reaching the stomach. So that in the case of fish its smell and repulsive taste can be masked (Khan *et al.*, 2019). Active ingredients used in food processing sector like fatty acids, vitamins are Nano encapsulated to improving stability, taste and color. Mainly in preservation of meat, sausage and cheese industry additives are nano encapsulated to accelerate its production. Nano droplets spray is another invention in food processing field. Droplets size range of about 87nm are used for vitamin B12 supplements(Berger 2019). Nanoemulsion with a size range of 10 to 100 nm are used in food processing (Acosta 2009). Nanoemulsions are formulated with dispersing oil/water phase in a continuous water/oil phase (Oberdorster *et al.*, 2007). Examples of nano emulsified food products are sweeteners, salad dressing, flavored oil etc. core material in nanoemulsion are released in favor of environmental conditions like temperature, pH etc. (Kumar, 2000).

(ii) Food Packaging

Food packaging is an important sector which help in retaining nutrition, quality and safety of food product during handling and storage. Packaging is the third largest industry in the world with a GNP of 2% in developed countries (Han 2015). Packaging is an important factor for meeting market demands and consumer preferences that determine the successfulness of a new food product. The role of packaging in food products is to protection from surrounding, to contain the food, to extend shelf life of food product and improve quality of product (Cole, 2006). Application of nanotechnology can improve the food packaging characteristics like water resistance, mechanical and thermal stability and increasing shelf life by integrating plant extracts, enzymes and bacteriostatic agents (Berger, 2019). Green approach in food packaging is by the development of biodegradable packaging materials. Nanotechnology can be applied for sectors like improved, active and intelligent packaging. Most of the packaging materials are plastic polymers which will not degrade and cause environmental problems. Renewable, biodegradable, plant based and edible packaging materials will not harm the environment. Organic nano composite materialused are based on green tree extract, milk thistle extracts etc. In inorganic materials nano-clay, nanotubes nanoparticles are used. Problems associated with plastic packing of beer like flavor loss and spoilage can be avoided by usage of nano composite materials and it also reduce packaging waste.Biodegradable nano composites also offer enhanced properties like appearance, flavor and odor (Othman 2014). Smart packaging materials can be developed using nanoparticles. Silicate nanoparticles are used in packaging materials to extend shelf life of product (Bayer *et al.*, 2005). To improve barrier resistance of the packaging material nanoparticles silver, magnesium, silicate and zinc oxides are used (Avella *et al.*, 2005). Mechanical and heat resistance properties of packaging materials can be improved using nanoparticles by changing permeability of material. Polymer silicate nano composite material provide good mechanical strength, gas barrier and heat resistant properties than that of polymer based packaging materials (Brody, 2006). To increase toughness of materials nanocomposites are firstly prepared by combining organic substances like protein, carbohydrates or lipids with inorganic components like calcium carbonate. One of which is potato starch and calcium carbonate based nanocomposites they are more thermostable and biodegradable. These nanocomposites materials can replace polystyrene used for food packaging (Moraru *et al.*, 2003). Other examples of nanocomposites are produced from starch and polybetahydroxyoctanoate, starch and tucinin whiskers, starch and clay nano composite (Maynard and Kuempel2005, Othman, 2014). Nanocomposites are also used as coating material for plastic films to improve its performance (Moore, 1999). Natural materials like smectic clays and montmorillonite are used for making nanomaterials (Quarmley *et al.*, 2001). Nanocomposites made with 3–5% montmorillonite improves thermal stability and enhanced barrier property against volatiles, oxygen, moisture and carbon dioxide. Ethylene vinyl alcohol and poly lactic acid polymer based nano clays found to increase barrier properties (Lagaron *et al.*, 2005).

Nanolaminatesis another innovation in food packaging where it can reduce food loss from spoilage. Nanolaminates consists of two or more nanomaterials are physically or chemically bonded to each other. In Nano lamination food product is coated with thin layer of nanolaminates mainly made of edible protein, polysaccharides and lipids. Nanolaminates improve appearance, color, texture and taste of food products (Sharon *et al.*, 2010). Nanolaminates are applied in the form of edible coating, film and foams in meat, chocolate, vegetable and baking industries. Nanolaminates based coating improve barrier properties to moisture, lipid or gases and also enhance textural properties of food. Nanolaminates based coating can be used as a carrier for flavor, color, nutrients, antimicrobials, and antioxidants (Ponce *et al.*, 2008, The *et al.*, 2008). Common raw materials used for making nanolaminates for coating/films are of polysaccharides, proteins or lipids. Advantage of edible coating based on nanolaminates are of fragility and extremely thin nature of coating material (Kotov 2003). Nano composites are another type of packaging material in which different polymers are combined to increase the overall activity of packaging material (Yotova *et al.*, 2013). Nanocomposites packaging have high barrier property thus food stored fresh without microbial contamination for long period. In the case of carbonated beverages nanocomposites packaging prevent the release of co2. Nanocomposites prevent movement of metal ions to the food product. Commercially used nanocomposites are polylactic acid, gelatin and LDPE (Beigmohammadi *et al.*, 2016). Nanoclays are transparent, low density, ecofriendly packaging material having good flow properties. Imperm, Aegis and Urethan are the commercially available ones (Davis and Song 2006). Nanocore is another nanomaterial having gas barrier property used in plastic beer containers(He *et al.*, 2010). Titanium dioxide nanoparticle is used in packaging material to control*Escherichia coli* contamination (Momin *et al.*, 2013). It also helps in whitening of dairy products like milk and cheese (Acosta 2009). Silicon dioxide is another nanoparticle use in packaging material and function as anticaking agentby removing moisture from food (Zhao *et al.*, 2008). Silver nanoparticles do not possess any harmful effect and are approved by FDA in prescribed amount. They do not leach into the food product (Addo Ntim *et al.,* 2015).

B. Nutraceuticals and Functional Food Development

Nano technological approaches are used in the production of on demand foods in which nutrients are delivered to the cells as per requirement otherwise remain dormant inside body. More effective nutrient delivery as a response of body requirement is made possible in functional food development with the help of nanotechnology. Encapsulation is a technique used in food processing to control active material interaction in food matrix, mask taste and odor, availability of active material in time, protection of active material against environmental conditions like heat, moisture, degradation by chemical and biological components during processing and storage(Ubbink and Krüger2006, Weiss *et al.*, 2006). Nanocapsules can be prepared by different methods like emulsion diffusion, nanoprecipitation, droplet emulsion, emulsion coacervation, double emulsion, polymer coating and layer by layer method. Active components like vitamins, minerals, antioxidants, probiotics, colorants, flavors, micronutrients and antimicrobials are found to be nano encapsulated(Hsieh and Ofori 2007). Different types of delivery systems like emulsion, associated colloids, and biopolymer matrix are used to maintain activity and increase the shelf life of active components during storage(Jelinski 2002).

Nanoencapsulation is one method which is used for the production of these type of functional foods. Nano encapsulated probiotics are one of them. Probiotics are bacterial formulation which function in gastro intestinal tract for beneficial effect. Nano encapsulated probiotics possess the ability to respond to body signals like de-novo vaccines (Vidhyalakshmi *et al.*, 2009). Encapsulated nanoparticles are released inside the body in response to specific environmental triggers like change in solution condition or porosity or dissolution of particles (Weiss *et al.*, 2006). Targeted nano carriers are found to reduce toxicity and increase efficiency of distribution of active components(Khosravi-Darani *et al.,* 2007). Most convenient means of drug delivery system are through oral route. Nanotechnology based delivery system can control the bioactive component release inside body leading to better dosage pattern and it also lowers side effects. Thus, delivery technique based on nanotechnology is better than that of conventional delivery systems. These nanoencapsulation technique can protect the bioactive component from degradation by enzymes thus increases their bioavailability. Gastro intestinal tract have a natural ability to protect encapsulated materials this is advantages for oral delivery of nano encapsulated bioactive materials (Mir *et al.*, 2017). Dietary supplements can be prepared using nanotechnology, they contain active ingredients like minerals, amino acids, vitamins etc. Commercially available human nano supplements are vitamin A, D, E, C, riboflavin, folic acid and β-carotene, probiotics *Bifidobacterium* and *Lactobacillus* species, antioxidants Polyunsaturated fatty acids, quercetin, hydroxytyrosol, SeNPs, lycopene, resveratrol, lutein, polyphenols, curcumin, astaxanthin, catechins, and naringin. Nano encapsulation of edible oils can increase its stability by imparting protection against lipid oxidation. Thus, fatty acid rich oils can be nano encapsulated to increase its bioavailability and stability. Stability of oil in water emulsion can be improved using this technique. Ultrasonication and high-pressure homogenization is the technique used for the production of nanoemulsions. Nano delivery systems are classified into 3 based on type of molecule used for the process. 1) Protein based nano delivery systems: - proteins are amphilic in nature, nontoxic, high level of biocompatibility, they possess strong affinity to bioactive compounds via hydrogen bond, disulfide bond and hydrophobic interactions. For example, casein micelles in milk. 2) Polysaccharide based nano delivery system: - they are hydrophilic in nature, have glycosidic bond between monosaccharides, also contains many functional groups like amino, carboxyl and hydroxyl groups. For example, edible nanomaterial phytoglycogen, it is a naturally occurring nano material extracted from maize variety. It is used for encapsulating curcumin, nisin, quercetin(Chen and Yao 2017)and lutein. 3) Hybrid nano delivery system: - it is made with a combination of two or more biopolymers. They have combined benefits of each biopolymer used in the system and also possess colloidal properties. Example of protein-protein complex system is zein-caseinate nanoparticles (Luo 2020). Curcumin, active component of turmeric its encapsulation using nanoemulsions are found to retain its color and health benefits. Study showed that curcumin nanoemulsion improve its oral bioavailability and stability of curcumin (Wang *et al.*, 2008) and chitosan based nanoparticles improve bioavailability of tea polyphenols catechins and epigalocatechin gallate. Reconstituted casein micelles from sodium caseinate and salt is used to encapsulate bioactive compounds. Conventional emulsions are replaced with nano emulsion due to its higher adsorption rate due to increased surface area a result of smaller size. Thus, making nanoemulsions more efficient and widely acceptable.Commercially used nanomaterials for delivery system are carbon nanotubes, gold nanoparticles, quantum dots, magnetic nanoparticles, dendrimers, silicon nanomaterials, graphene nanomaterials and conducting polymers. Gold nanoparticles are frequently used nanomaterial for delivery system, surfactants need to add with them to avoid aggregation. Quantum dots have size less than 10 nm and have luminance property and are highly soluble. Magnetic nanoparticles have size range of 50 to 500 nm and are used for making biosensors. Graphene based nanomaterials are used fabrication ofbiosensors which convert targeted molecules into detectable measurements. Silicone nanomaterials are used in bio imaging techniques. Dendrimers are complex shaped particle having size range of 2 to 20 nm (Lagaron *et al*., 2005). Dendrimers is a class of polymer used as a coating material in delivery system. They have regular, branched 3D structure making it suitable to use in sensors, catalysis and drug delivery system(Hughes 2005). Features of dendrimers are nontoxicity, non-immunogenicity and biodegradability(Khosravi- Darani *et al*., 2007). Cochleates are another type of delivery system with a structural characteristic of continuous solid with large multiple lipid layer sheet rolled into a spiral geometry(Mozafari *et al*., 2006). Outer layer of cochleates combine with cell membrane releasing contents into the cell. Because of the strong structure active materials are prevented from deterioration and is more effective system of oral intake (Zarif 2003). Cochleates are effective for functional food development (Moraru *et al*., 2003). It is widely used for nutraceuticals and antimicrobial delivery (Weiss *et al*., 2006). Active compounds like proteins, drugs, peptides, large hydrophilic compounds and compounds with poor water solubility can be encapsulated using cochleates (Gould-Fogerite *et al*., 2003). Colloids are also applied in encapsulation technique for the delivery of amphilic, polar and non-polar functional particles(Flanagan and Singh 2006). Micellets with a diameter of 5-10nm are used for encapsulating non polar molecules like vitamins, antimicrobials, lipids, antioxidants and flavors (Weiss and McClements 2002). Novasol CT is a nanoparticle solution used for applying antioxidants in food and beverages. Micelles are the nanoparticle present in novosal CT which carry antioxidants and vitamins into food and beverages. Nanoemulsion prepared using high pressure homogenizer or micro fluidizer have particle size in the range of 100 to 500 nm (McClements 2005). Much smaller particle size can be achieved using ultrasonication method. Active ingredients are incorporated inside the droplet, in continuous phase or interfacial region to avoid degradation during processing and storage (Weiss and McClements, 2002). Multilayer emulsion or multiple emulsion is a recent discovery in nano emulsification technique which possess novel encapsulation and delivery properties. Oil in water in oil or water in oil in water are the method of emulsification(Garti *et al.*, 2005). In multilayer emulsion functional ingredients like proteins, polysaccharides or phospholipids are trapped inside the core and surrounded by nano-size multiple layers of different poly electrolytes,thus they have more stability against environmental stress than that of conventional single layer emulsion (Guzey and McClements 2006). Lipid vesicles are suitable delivery system for broad range of substance sin functional food (Taylor *et al.*, 2005). Liposomes are closed two-layer structure consisting of lipid and phospholipid molecules (Khosravi-Darani *et al.*, 2007) and it can be easily prepared by heating without any harmful procedure or chemical involved. Lipid vesicles have uni or multi lamellar structure with one or more bilayer shells (Mortazavi *et al.*, 2007). Liposomes are applied for the controlled delivery of enzymes, vitamins and flavors in food industry (Taylor *et al.*, 2005). For cheese manufacturing industry liposomes are used for encapsulating enzymes to increase speed of ripening (Law and King, 1985). In diary industry liposomes are used to encapsulate vitamins for enhancing nutritional quality of products (Banville *et al.*, 2000). Studies showed that liposomes can protect bioactive ingredient from degradation during processing. Forexample, liposome encapsulated retinol degradation was reduced with the addition of vitamin E (Lee and Martin 2002). Arachaesomes are type of thermostable lipids used as nanoparticle for the delivery of antioxidants (Alfadul and Elneshwy, 2010). Other nanocarriers for delivery of active materials are nano gels, silica nano particles, nano fibers, nano sponges, cyclodextrin complexes and layered double hydroxides.

C. Nanosensors

Nanosensors are used for food safety to detect deterioration in food. Freshness and shelf life of food products can be monitored using nanosensors. Other than food safety nanosensors have application in food processing and production. Changes and spoilage in food product quality over time can be detected using nanosensors. Portable nanosensors have ability to detect toxins and pathogens in real time in food (Handford *et al.*, 2014). Nanosensors detect food spoilage by sensing gas released by food during spoilage. Major gases detected by nanosensors are ammonia, nitrogen oxide, hydrogen sulfide and hydrogen (Chun *et al.*, 2010). Nanosensors can be applied in grain storage system which detect spoilage in grain and also change in food storage environment to ensure grain quality. Nanosensors are made with thousands of nanoparticles which detect insect infestation and even fungus spores in grain storage system.

Features of nanosensors are they are tiny in size, less weight and low powder requirement making it suitable for keeping inside grain storage system. Nano sensor strips can detect spoilage in food by sensing gas released from food showing color change in strip (Pradhan *et al.*, 2015). By the application of nanotechnology effective sensors can be developed to detect microbes, spoilage, gases and contaminants even in packaged foods and also storage system.Apart from detection of microbes and gases toxins in packaged foods can also be detected using nanobiosensors. Nanobiosensors are reported to have potential application in detecting safety status of foods during processing, storage and distribution of food products(Helmke and Minerick2006). An electrochemical nanobiosensors for glucose detection is fabricated by layer-by- layer assembly of polyelectrolytes (Rivas *et al.*, 2007). In chocolate manufacturing industry for the detection of peanut allergic proteins and pathogens liposome nanovesicles are used (Edwards and Baeumner 2006, Wen *et al.*, 2005). Microbes like *E.coli, Salmonella* and *Listeria monocytogenes* can be detected using assay of G-liposomal protein nanovesicle and immunogenic bead (Cheng *et al.*, 2006). Electronic nose or E nose is another invention in biosensor for identification of odors. Main component of E nose is a nanosensors for gas detection for example zinc oxide nanowires (Hossain *et al.*, 2005). Poly dimethyl siloxane chip supported by fluid bilayer membraneattached with antibodies to enterotoxin are used for the detection of Staphylococcus enterotoxin B in milk (Dong *et al.*, 2006). For detection of DNA quartz crystal biosensor fitted with gold nano particle of 50 nm have been used (Zhao *et al.*, 2001). Nanoelctronic devices are developed with gold array having 1- Dodecanethiolencapsulated colloidal particles (Ge *et al.*, 2003). Studies found out that thiol modified DNA probe can be immobilized in to gold nanoparticle coated electrode to detect complementary DNA sequences(Liu *et al.*, 2005). DNA sensor coated with gold nano particles can detect *E.coli* metabolites. These nanoparticles amplify the signal thus increasing the limit of detection of microbes (Mao *et al.*, 2006). Carbon nanotubes are nanomaterials used for fabrication of nanosensors. Carbon nanotubes are made by coiling graphite sheets in honeycomb structure (Moraru *et al.*, 2003). These types of biosensors are used for the detection of proteins, enzymes, antibodies and DNA(Lee and Martin 2002).

6.4 Safety Aspects of Application of Nanotechnology in Food Sector

Not so many studies are available on the effect of nanomaterials inside the human body. Presence of nonedible nanoparticles are a result of contact with nano packaging material. Nanoparticles may migrate from packaging material in to the food. Research show that nanoparticle entered in to body through oral means will pass through intestine and it get eliminated from the body rapidly.

Nanoparticles toxicity inside the human body is depended upon factors like material property, route of entrance, susceptibility and state of individual. Other ways of nanomaterials into the body are through inhalation and penetration through skin especially for workers in factors which manufacture nanomaterials. Some studies revealed that nanoparticle cannot penetrate healthy intact skin. Studies shown that if the nano particles are hydrophilic in nature with positively charged surface its circulation time increases drastically. To analyze the toxicity of nano materials it is necessary to understand the physico chemical characteristics of nanomaterials, its surface chemistry, and mode of preparation, state of agglomeration and source of material. Another possible adverse effect of nanoparticles are carcinogenicity and genotoxicity. Zinc oxide nano particles found to have genotoxic potential in human epidermal cell but large zinc oxide particles don't have this proves the impact of particle diameter. High dose of nano particles asbestos is found to possess carcinogenic effect (Dimitrijevic *et al.*, 2015).

6.5 Regulations and Laws Regarding Application of Nano Technology in Food Sector

The Food and Drug Administration (FDA) regulates nanoparticles use on the basis of products (FDA 2004, Weiss *et al.*, 2006). FDA has not focused on processing techniques based on nanotechnology but have given regulation regarding products with nanosize range materials. There are other government agencies focusing on application of nanotechnology reduce environmental issues and to treat diseases (FDA 2004). If nanoparticles are used as additives in food it should be labelled with – and also have an E numbering system as suggested by Institute of Food Science and Technology (IFST) (Maynard and Kuempel 2005). British government have accepted this suggestion and these ingredients are subjected to full safety assessment before using in food products(Weiss *et al.*, 2006). Codex Alimentarius commission created in 1963 have a food code containing all the standards, guidelines and codes of practice for handling food products. As per the recommendation of World Health Organization and Food and Agricultural Organization, Codex Alimentarius updates the practice of nanotechnology in food and agriculture. FAO and WHO conducted an expert consultation in 2008 with a scope of nanotechnologies with application in the food and agriculture sectors and their potential impact on food safety along the entire food chain. Risk assessment analysis is required before applying any food laws to nano foods. Organizations like International Standards for Organization (ISO) and ASTM International on terminology, nomenclature, measurement and characterization, and environment, safety, and health are expected to develop standards for nanotechnology.

6.6 Summary

Nanotechnology have innovatory application in food processing and preservation sector. Special safety requirements are necessary before applying it to food products. Nanotechnology is a highly growing and trending field which need more in-depth cell line studies while applying for direct food products. Biodegradable, nontoxic nano products bring green aspects to food processing. Researches have to focus more on naturally occurring nanoparticles like phytoglycogen as carrier system for targeted delivery of bioactive components rather than chemicals. As the size of nano materials are very low risk associated with them is high. Strong laws and regulations are required for the usage of nanomaterials in food processing.

6.7 References

Acosta E. 2009. Bioavailability of nanoparticles in nutrient and nutraceutical delivery. Current opinion in colloid & interface science, 14(1), 3-15.

Addo Ntim S, Thomas TA, Begley TH, Noonan GO. 2015. Characterisation and potential migration of silver nanoparticles from commercially available polymeric food contact materials. Food Additives & Contaminants: Part A, 32(6), 1003-11

Aguilera JM. 2005. Why food microstructure? Journal of food engineering, 67(1-2), 3-11.

Alfadul SM, Elneshwy AA. 2010. Use of nanotechnology in food processing, packaging and safety–review. African Journal of Food, Agriculture, Nutrition and Development, 10(6).

Avella M, De Vlieger JJ, Errico ME, Fischer S, Vacca P, Volpe MG. 2005. Biodegradable starch/clay nanocomposite films for food packaging applications. Food Chemistry, 93(3), 467-74.

Bai YX, Li YF, Yang Y, Yi LX. 2006. Covalent immobilization of triacylglycerol lipase onto functionalized nanoscale SiO2 spheres. Process Biochemistry, 41(4), 770-7.

Banville C, Vuillemard JC, Lacroix C. 2000. Comparison of different methods for fortifying Cheddar cheese with vitamin D. International Dairy Journal, 10(5-6), 375-82.

Bayer J, Rädler JO, Blossey R. Chains, dimers, and sandwiches: 2005. Melting behavior of DNA nanoassemblies. Nano letters, 5(3), 497-501.

Beigmohammadi F, Peighambardoust SH, Hesari J, Azadmard-Damirchi S, Peighambardoust SJ, Khosrowshahi NK. 2016. Antibacterial properties of LDPE nanocomposite films in packaging of UF cheese. LWT-Food Science and Technology, 65, 106-11.

Berger M. Nanoengineering: 2019. The Skills and Tools Making Technology Invisible. Royal Society of Chemistry.

Brody AL. 2006. Nano and food packaging technologies converge. Food technology (Chicago), 60(3), 92-4.

Cheng MM, Cuda G, Bunimovich YL, Gaspari M, Heath JR, Hill HD, Mirkin CA, Nijdam AJ, Terracciano R, Thundat T, Ferrari M. 2006. Nanotechnologies for biomolecular detection and medical diagnostics. Current Opinion in Chemical Biology, 10(1), 11-9.

Chen, H., & Yao, Y. 2017. Phytoglycogen improves the water solubility and Caco-2 monolayer permeation of quercetin. Food chemistry, 221, 248-257.

Chun JY, Kang HK, Jeong L, Kang YO, Oh JE, Yeo IS, Jung SY, Park WH, Min BM. 2010. Epidermal cellular response to poly (vinyl alcohol) nanofibers containing silver nanoparticles. Colloids and Surfaces B: Biointerfaces, 78(2), 334-42.

Cole MF, Bergeson LL. 2006. FDA regulation of food packaging produced using nanotechnology. Foodsafety magazine.

Davis G, Song JH. 2006. Biodegradable packaging based on raw materials from crops and their impact on waste management. Industrial Crops and Products, 23(2), 147-61.

Dimitrijevic M, Karabasil N, Boskovic M, Teodorovic V, Vasilev D, Djordjevic V, Kilibarda N, Cobanovic N. 2015. Safety aspects of nanotechnology applications in food packaging. Procedia Food Science, 5, 57-60.

Dong Y, Phillips KS, Cheng Q. 2006. Immunosensing of Staphylococcus enterotoxin B (SEB) in milk with PDMS microfluidic systems using reinforced supported bilayer membranes (r-SBMs). Lab on a Chip, 6(5), 675-81.

Edwards KA, Baeumner AJ. 2006. Liposomes in analyses. Talanta, 68(5), 1421-31.

Flanagan J, Singh H. 2006. Microemulsions: a potential delivery system for bioactives in food. Critical reviews in food science and nutrition, 46(3), 221-37

Food and Drug Administration. 2004. FDA Regulation of Nanotechnology Products. Available at http: //www.fda.gov/nanotechnology/regulation.html (accessed 2004/03/05)

Garti N, Spernath A, Aserin A, Lutz R. 2005. Nano-sized self-assemblies of nonionic surfactants as solubilization reservoirs and microreactors for food systems. Soft Matter, 1(3), 206-18.

Ge C, Liao J, Wang Y, Chen K, Gu N. 2003. DNA assembly on 2-dimensional array of colloidal gold. Biomedical Microdevices, 5(2), 157-62.

Ghouri MZ, Khan Z, Khan SH, Ismail M, Aftab SO, Sultan Q, Ahmad A. 2020. Nanotechnology: Transformation of agriculture and food security. Bioscience, 3:19.

Giraldo JP, Wu H, Newkirk GM, Kruss S. 2019. Nanobiotechnology approaches for engineering smart plant sensors. Nature nanotechnology, 14(6), 541-53.

Gould-Fogerite S, Mannino RJ, Margolis D. 203. Cochleate delivery vehicles: applications to gene therapy. Drug Delivery Technol, 3(2), 40-7.

Grumezescu A, Oprea AE, editors. 2017. Nanotechnology applications in food: flavor, stability, nutrition and safety. Academic Press.

Guzey D, McClements DJ. 2006. Formation, stability and properties of multilayer emulsions for application in the food industry. Advances in Colloid and Interface Science, 128, 227-48.

Han JH. 2015. New technologies in food packaging: overview. In: Han JH (ed) Innovations in food packaging. Elsevier Academic Press, London, pp 3–11.

Handford CE, Dean M, Henchion M, Spence M, Elliott CT, Campbell K. 2014. Implications of nanotechnology for the agri-food industry: opportunities, benefits and risks. Trends in Food Science & Technology, 40(2), 226-41

He CX, He ZG, Gao JQ. 2010. Microemulsions as drug delivery systems to improve the solubility and the bioavailability of poorly water-soluble drugs. Expert Opinion on Drug Delivery, 7(4), 445-60.

Helmke BP, Minerick AR. 2006. Designing a nano-interface in a microfluidic chip to probe living cells: challenges and perspectives. Proceedings of the National Academy of Sciences, 103(17), 6419-24

Hossain MK, Ghosh SC, Boontongkong Y, Thanachayanont C, Dutta J. 2005. Growth of zinc oxide nanowires and nanobelts for gas sensing applications. InJournal of Metastable and Nanocrystalline Materials, 23, 27-30.

Hsieh PY, Ofori JA. 2007. Innovations in food technology for health. Asia Pacific Journal of Clinical Nutrition, 16(S1), 65-73.

Hughes GA. 2005. Nanostructure-mediated drug delivery. Nanomedicine: Nanotechnology, Biology and Medicine, 1(1), 22-30.

Hwang ET, Gu MB. 2013. Enzyme stabilization by nano/microsized hybrid materials. Engineering in Life Sciences, 13(1), 49-61.

Institute of Food Science and Technology. 2006. Nanotechnology Information Statement. (IFST) Trust Fund, London, UK. Available at: http: //www.ifst.org/nano.pdf, 2006

Jelinski, L. 2002. Biologically related aspects of nanoparticles, nanostructured materials, and nanodevices. In: Nanostructure Science & Technology. Siegel, R.W., Hu, E., and Roco, M.C., Eds. A worldwide study, prepared under the guidance of National Science and Technology Council and the Interagency Working Group on NanoScience, Engineering, and Technology, 2002.

Khan Z, Khan SH, Ghouri MZ, Shahzadi H, Arshad SF, Waheed U, Firdous S, Ahmad A. 2019. Nanotechnology: An Elixir to Life Sciences.

Khosravi-Darani K, Pardakhty A, Honarpisheh H, Rao VM, Mozafari MR. 2007. The role of high- resolution imaging in the evaluation of nanosystems for bioactive encapsulation and targeted nanotherapy. Micron, 38(8), 804-18.

Kotov NA. 2003. Layer-by-layer assembly of nanoparticles and nanocolloids: intermolecular interactions, structure and materials perspectives. Multilayer Thin Films, 207, 269.

Kumar MN. 2000.A review of chitin and chitosan applications. Reactive and functional polymers, 46(1), 1-27.

Kumar V, Khan I, Gupta U. 2020. Lipid-dendrimer nanohybrid system or dendrosomes: evidences of enhanced encapsulation, solubilization, cellular uptake and cytotoxicity of bortezomib. Applied Nanoscience, 24:1-4.

Lagaron JM, Cabedo L, Cava D, Feijoo JL, Gavara R, Gimenez E. 2005. Improving packaged food quality and safety. Part 2: Nanocomposites. Food Additives and Contaminants, 22(10), 994-8.

Law BA, King JS. 1985. Use of liposomes for proteinase addition to Cheddar cheese. J. Dairy Res, 52, 183-8.

Lee SB, Martin CR. 2002. Electromodulated molecular transport in gold-nanotube membranes. Journal of the American Chemical Society, 124(40), 11850-1.

Liu SF, Li YF, Li JR, Jiang L. 2005. Enhancement of DNA immobilization and hybridization on gold electrode modified by nanogold aggregates. Biosensors and Bioelectronics, 21(5), 789-95.

Luo Y. 2020. Perspectives on Important Considerations in Designing Nanoparticles for Oral Delivery Applications in Food. Journal of Agriculture and Food Research, 100031.

Mao X, Yang L, Su XL, Li Y. 2006. A nanoparticle amplification-based quartz crystal microbalance DNA sensor for detection of Escherichia coli O157: H7. Biosensors and Bioelectronics, 21(7), 1178-85.

Mathew AP, and Dufresne A. 2002. Morphological investigation of nanocomposites from sorbitol plasticized starch and tunicin whiskers. Biomacromolecules, 3(3), 609-17, 2002

Maynard AD, Kuempel ED. 2005. Airborne nanostructured particles and occupational health. Journal of nanoparticle research, 7(6), 587-614

McClements DJ, Decker EA. 2000.Lipid oxidation in oil-in-water emulsions: Impact of molecular environment on chemical reactions in heterogeneous food systems. Journal of food science, 65(8), 1270-82.

McClements DJ. 2005. Food emulsions: principles, practices, and techniques, pp 609, CRC press.

Mir SA, Shah MA, Mir MM, Iqbal U. 2017. New horizons of nanotechnology in agriculture and food processing industry. InIntegrating Biologically-Inspired Nanotechnology into Medical Practice, pp. 230-258, IGI Global.

Momin JK, Jayakumar C, Prajapati JB. 2013. Potential of nanotechnology in functional foods. Emirates Journal of Food & Agriculture (EJFA), 25(1).

Moore S. 1999. Nanocomposite achieves exceptional barrier in films. Modern Plastics, 76(2), 31–32.

Moraru CI, Panchapakesan CP, Huang Q, Takhistov P, Liu S, Kokini JL. 2003. Nanotechnology: a new frontier in food science, food Technol, 57(12), 24-29.

Mortazavi SM, Mohammadabadi MR, Khosravi-Darani K, Mozafari MR. 2007. Preparation of liposomal gene therapy vectors by a scalable method without using volatile solvents or detergents. Journal of biotechnology, 129(4), 604-13.

Mozafari MR, Flanagan J, Matia-Merino L, Awati A, Omri A, Suntres ZE, Singh H. 2006. Recent trends in the lipid-based nanoencapsulation of antioxidants and their role in foods. Journal of the Science of Food and Agriculture, 86(13), 2038-45.

Newsome R. 2007. Codex vital in global harmonization. Food technology. Institute of Food Technologists, IFT, 61: 10, Oct 2007. Available at: http: //www.ift.org/.

Oberdörster G, Stone V, Donaldson K. 2007. Toxicology of nanoparticles: a historical perspective. Nanotoxicology, 1(1), 2-5.

Omanović- Mikličanina E, Maksimović M. 2016.Nanosensors applications in agriculture and food industry. Bull Chem Technol Bosnia Herzegovina, 47, 59-70.

Othman SH. 2014. Bio-nanocomposite materials for food packaging applications: types of biopolymer and nano-sized filler. Agriculture and Agricultural Science Procedia, 2, 296-303.

Park DP, Kim JW, Liu F, Choi HJ, Joo J. 2003. Synthesis and characterization of polypyrrole/ organoclay nanocomposite. Synthetic metals, 135, 713-4

Ponce AG, Roura SI, del Valle CE, Moreira MR. 2008. Antimicrobial and antioxidant activities of edible coatings enriched with natural plant extracts: in vitro and in vivo studies. Postharvest biology and technology, 49(2), 294-300.

Pradhan N, Singh S, Ojha N, Shrivastava A, Barla A, Rai V, Bose S. 2015. Facets of nanotechnology as seen in food processing, packaging, and preservation industry. BioMed research international.

Quarmley J, Rossi A. 2001. Nanoclays. Opportunities in polymer compounds. Industrial Minerals, 400, 47-9.

Rahmati F, Hosseini SS, Safai SM, Lajayer BA, Hatami M. 2020. New insights into the role of nanotechnology in microbial food safety. 3 Biotech, 10(10), 1-5.

Ranjan S, Dasgupta N, Chakraborty AR, Samuel SM, Ramalingam C, Shanker R, Kumar A. 2014. Nanoscience and nanotechnologies in food industries: opportunities and research trends. Journal of nanoparticle research, 16(6), 2464.

Rivas GA, Miscoria SA, Desbrieres J, Barrera GD. 2007. New biosensing platforms based on the layer- by-layer self-assembling of polyelectrolytes on Nafion/carbon nanotubes-coated glassy carbon electrodes. Talanta, 71(1), 270-5

Sharon M, Choudhary AK, Kumar R. 2010. Nanotechnology in agricultural diseases and food safety. Journal of Phytology.

Singh T, Shukla S, Kumar P, Wahla V, Bajpai VK, Rather IA. 2017. Application of nanotechnology in food science: perception and overview. Frontiers in microbiology, 8, 1501.

Srivastava AK, Dev A, Karmakar S. 2018. Nanosensors and nanobiosensors in food and agriculture. Environmental chemistry letters, 16(1), 161-82.

Taylor TM, Weiss J, Davidson PM, Bruce BD. 2005. Liposomal nanocapsules in food science and agriculture. Critical reviews in food science and nutrition, 45(7-8), 587-605.

The DP, Debeaufort F, Luu D, Voilley A. 2008. Moisture barrier, wetting and mechanical properties of shellac/agar or shellac/cassava starch bilayer bio-membrane for food applications. Journal of Membrane Science, 325(1), 277-83.

Ubbink J, Krüger J. 2006. Physical approaches for the delivery of active ingredients in foods. Trends in Food Science & Technology, 17(5), 244-54

Verma ML, Barrow CJ, Puri M. 2013. Nanobiotechnology as a novel paradigm for enzyme immobilisation and stabilisation with potential applications in biodiesel production. Applied Microbiology and Biotechnology, 97(1), 23-39.

Vidhyalakshmi R, Bhakyaraj R, Subhasree RS. 2009. Encapsulation —the future of probiotics-a review. Adv Biol Res, 3(3-4), 96-103.

Wang X, Jiang Y, Wang YW, Huang MT, Ho CT, Huang Q. 2008. Enhancing anti-inflammation activity of curcumin through O/W nanoemulsions. Food Chemistry, 108(2), 419-24.

Weiss J, McClements DJ. 2002.Mass transport phenomena in emulsion containing surfactants. In —Encyclopedia of Surface and Colloid Science. ed. P. Somasundaran and A. Hubbard.

Weiss J, Takhistov P, McClements DJ. 2006. Functional materials in food nanotechnology. Journal of food science, 71(9), R107-16.

Wen HW, Borejsza-Wysocki W, DeCory TR, Baeumner AJ, Durst RA. 2005. A novel extraction method for peanut allergenic proteins in chocolate and their detection by a liposome-based lateral flow assay. European Food Research and Technology, 221(3-4), 564-9.

Yotova L, Yaneva S, Marinkova D. 2013. Biomimetic nanosensors for determination of toxic compounds in food and agricultural products. Journal of Chemical Technology & Metallurgy, 48(3).

Yu L, Banerjee IA, Gao X, Nuraje N, Matsui H. 2005. Fabrication and application of enzyme-incorporated peptide nanotubes. Bioconjugate chemistry, 16(6), 1484-7.

Zarif L. 2003. Nanocochleate cylinders for oral & parenteral delivery of drugs. InJournal of liposome research, Vol. 13, No. 1, pp. 109-110, 270 Madison AVE, New Yourk, NY 10016 USA: Marcel Dekker INC, 2003

Zhao HQ, Lin L, Li JR, Tang JA, Duan MX, Jiang L. 2001. DNA biosensor with high sensitivity amplified by gold nanoparticles. Journal of Nanoparticle Research, 3(4), 321-3.

Zhao R, Torley P, Halley PJ. 2008. Emerging biodegradable materials: starch-and protein-based bio- nanocomposites. Journal of Materials Science, 43(9), 3058-71.

7

Valorization of By Products in Agro Commodity Processing Sectors

Jayati Pal Chattopadhyay **and** ***Aditi Roy Chowdhury***

ABSTRACT

Approximately 30 % of fruits and vegetables are getting wasted each year in country like India. Not only that, there is a huge loss of agro based commodity during industrial processing and will be considered as industry generated waste. Utilization of such unexplored nutrients from agro processing sectors for making a sustainable food system will be a healthy approach to bring food security and nutrition to all of us. With this aim the present work addresses the utilization of jackfruit seeds, an agro-waste, to value added products. Jackfruit seed flour has a great potential in new food formulation along with wheat flour. The 15% (w/w) blending of jackfruit seed flour with wheat flour could be used in bread and the substituted bread is protein and fibre enriched. This attempt in developing such bakery items with blended flours could address the scarcity problem of non-wheat producing countries. Present study also looks into the alternate source of starch from spice processing industries. The nutrient starch could be explored for isolation from a nonconventional source from turmeric processing sector. The isolated starch could also be modified to enhance some of its functional properties and may be utilized in preparing value added product with a better acceptability and nutrition.Thus, the isolation of nutrients and value-added products out of fruits and vegetables and spice-processing waste would be a stepping stone to an initiative to zero hunger.

Keywords: Valorization, agro waste, jackfruit seed, bread, turmeric starch, modified starch, functional properties

Part-A: Introduction

7.1 Valorisation and Its Implications

Valorization means a measure to establish and maintain the price of a commodity with an assigned value or rather means a measure to establish a higher valued

*Corresponding Author

product. Sustainable management in food-processing originates new concept of valorization imparting commercial values to industrial residues. By-products, wastes or effluents can be utilized in two ways. Either they can be processed to recover the functional biological components or to transform them to precious metabolites by suitable technology. In these contexts, the overall processing should render a concept mapping with respect to making a sustainable world. Green processing rather green technology and related research has come up right at this juncture of time and getting much more attention. The blended technological concept with conventional technologies along with green processing may bring economically nutritionally environmentally friendly world.

Transformation of products sourced from agriculture, forestry or fisheries to final value-added product is the key concept of agro-processing industry. Either raw materials or derived intermediates from agro-sector are the major resources for agro-processing. Due to lack of proper storage and handling and infrastructure support in process technology, post process handling and supply chain and proper marketing, a huge part of agro based commodities are getting wasted in India and worldwide. Increasingly, industrial ecology concepts for eco-innovation, aim at zero waste economy whereby wastes are used as raw materials for new products and applications.

7.2 Concept of Food Waste

Around 1.3 billion metric tons of food, which is one-third of the global production, is being lost and wasted every year as per the report of United Nations Food and Agriculture Organization (FAO) (FAO 2014). Horticultural commodities contribute the most (upto 60%) in this loss and waste among all types of foods (Gustavsson *et al.* 2011). Large scale of food production and supply system demand efficient food storage and handling guided by suitable regulatory bodies (HLPE, 2014).

Presently, value-addition to the agricultural products gain much attention in transforming agro- products to end-use format. 98% of agricultural production in developed countries undergoes industrial processing, whereas the contribution to processing is much less in developing countries. Developing countries generate 40% to 60% of manufacturing value added and agro industrial products. In developed countries, 42% of food waste is produced by households, while 39% losses occur in the food manufacturing industry, 14% in food service sector and remaining 5% in retail and distribution. Many cereals, such as, sorghum, cassava, rice and many more are worth to be processed and get exported.

Assessment of loss can be conducted either in quantitative or in qualitative manner (FAO 2014). Loss in volume or mass directly indicates reduction in accessibility of food for consumption, which is quantitative measure. Decrease in edibility, nutrition, caloric value, consumer acceptability and economic value represent qualitative loss.

An approximate estimated loss of one-third of worldwide production of fruits and vegetables, not reachable to the consumer, has been given by Kader (2005). In developed countries, loss is also significant even at the consumer end. An overall average of 12% postharvest loss in the United States are estimated (Kader 2005), whereas, for United Kingdom, the value issuggested to be 9% (Garnett, 2006). Buzby and others (2011) also estimated that per capita loss of fruits and vegetables at retail and consumer level was almost141 $. FAO (2014) revealed that in China, India, Philippines, and the United States, significant percentage of fruit and vegetable waste is generated during processing as well as in supply-chain of fruits and vegetables and products based on these produce. This might be due to an imbalance between extra food supply and its demand, as identified by Kummu and others (2012). So, suitable processing, supply-chain and distribution network of raw fruits and vegetables and their value-added products could avert the situation of 47% agricultural loss and 86% consumption waste.

7.3 Concept of Green Technology

The basis of this technology is to build up a process, which is completely sustainable in nature. It involves techniques of green chemistry, bio-based components, soil-to-soil concept, environmental monitoring etc. to protect the natural eco-system of environment and is popularly known as clean technology. Some major benefits, associated with it, are recycling of waste, purification of air and water and rejuvenating ecosystems. The main goal is to conserve nature, and to remedy the negative impact that humans have on it.

(i) Green Technology in Managing Agro-waste

Agro-waste is the waste generated from agro-based materials during agricultural process, during post- harvest processing, in supply-chain and also during storage of agro-based raw materials. The process of collection, transport, disposal, recycling and monitoring of wastes is known as waste management. Recycling of waste is one of the techniques to manage this huge load of waste material other than land-filling, incineration, anaerobic digestion etc. It is defined as reprocessing waste materials into value-added products minimizing environmental load and keeps the environment clean. Globally, agricultural wastes increased at an average rate of 5%-10% per year. At present, around 350

MT are organic wastes from agricultural sources in India. Different ways are practiced to convert or recycle these agro-based wastes including production of bio-gas, alcohol, packaging materials, bio-fertilizer etc. and supply of raw materials in meat-processing, pulp/paper or leather industries. This, in turn, reduces overall load on environment and thus, substantiates with clean/green technology.

Case Study –I

7.4 Fruits and Vegetables Sector

(i) Waste Generation in Fruits and Vegetables Processing Sector

Fruits and vegetables processing industries generate huge waste which includes leaves, roots, tubers, skin, pulp, seeds, stones, pomace and many more (Othman, 2020). The discarded parts of fruits and vegetables processing industry become waste asa result of lack of proper handling operations and feasible technology in generating value added products or byproducts.

Generally, 25-30% of many fruits and vegetables are treated as waste materials having no further use. (Ajila *et al*. 2007). Total waste of fruits and vegetables during processing has been estimated by Laufenberg and others (2009).Almost 25% peels and seeds are generated with 58% of finished product during processing of mangoes (Ayala-Zavala *et al*. 2010; Joshi *et al*. 2012). Ratio of apple pulp with the by-product come in the range of 1:9 (approx) in an apple-pulp processing industry. Pineapple and pappya processing industry also yield about 9-14% of peel waste with 7% of seed and 9% of core material with 48-52% of final product. Large amount of solid waste (5-9MMT) is also contributed by grape and wine processing industries, which is about 20% to 30% of final processed products (Schieber and others 2001).

(ii) Value Added Components from Fruit and Vegetable Wastes and Socio-Economic Scenario

Scientists and researchers are coming up with innovations and innovative technologies to explore the feasibility of development of value-added products out of the discarded materials from fruit and vegetable processing sectors. Extraction, isolation and purification have been adopted to convert agro- waste into some bio based components like starch, pectins, fibres, proteins, bioactive components like polyphenolics, bio colours etc utilizing the generated wastes in various industries(Sagar *et al*.,2017).

The initiatives taken up by food processors, food manufacturer and relevant government support to impart green processing will surely contribute to nations GDP, employment creation, regional stability and sustainability and integration into global market. This utilization of food industry waste concept and sustainable green perspective would be a synergy between agro-industry and poverty alleviation in the country like India.

7.5 Research Based Study

A. Utilization of Fruit Processing Industry Waste through Fortification of Bread by Jackfruit Seed Flour

In India, some agricultural commodities are produced in plenty, but have limited use and are largely wasted. One such material is Jackfruit and its processing generates large amount of seed. Jackfruit seed is a material with good nutritional qualities. Attempts were made in this study to standardize the conditions for its use through bakery products which not only are consumed on daily basis but also the sector is a growing one where entrepreneurs can venture on a small scale at low cost.

The different aspects of getting the seeds in a usable and storable form and the extent of its inclusion along with the base wheat flour have been standardized considering different factors which might affect the acceptability. It may be concluded that it is certainly possible to enrich bread with jackfruit seed flour where the optimum blend is 15%. Industry could be benefitted from transforming the scientific knowledge to development of novel useful product and sustainable as well.

(i) Introduction

The major problems related to nutrition occurs due to inadequate supply of protein through normal diet and diet pattern. Among all, irrespective of ages, wheat flour and bakery products serve as staple or snack foods mostly in all parts of a developing nation like India , rural as well as urban sectors (Agrawal1990). Due to an increase in population and insufficient agricultural produce, thrust has been given in looking for an inexpensive high protein source which could improve and enhance the nutritional quality of the diet and would become a cheap source of it.

Jackfruit is treated as a well known and highly accepted fruit crop which is plenty in India, Bangladesh, Thailand and other parts of Tropics (Rahman *et al*, 1999, Burkill, 1997). The bulbs of ripe jackfruit are edible and consumed either fresh or processed. Different kinds of processed products are come up with value additions in India utilizing the ripe bulbs whereas the seeds are underutilized or under focused. Seeds make up around 10%-15% of the total fruit weight and have high carbohydrate and protein contents (Bobbio *et al*,

1978 and Kumar *et al*, 1988). As the seeds are normally discarded and usually considered as an agro waste, utilization of its flour would be worth exploring.

The seeds contain a lectin, Jacalin, which agglutinates erythrocytes (Chatterjee *et al.*, 1989), is highly specific for tumor associated Thomsen Friedenreich antigen (Mahanta *et al.*, 1990) and also has been reported to block HIV- 1 in vitro infection of lymphoid cells (Favero *et al*, 1993). It will be worthwhile to include jackfruit seed flour by blending with wheat flour. Hence the present study focuses on making nutritious bread utilizing jackfruit seed flour which can be a potent functional food adjunct.

It was estimated in India, total number of factories producing bread and biscuit which are under the category of medium and small-scale sectors is about 4800 (Tyagi S. K., 2006). Bread forms the major breakfast food which is around 50% of total bakery products. The production of bread and biscuit has increased from 5 to 19 lakh tonne from 1975-1990, to19 to 30 lakh tonne during 1990-1999, recording sixfold increase in a quarter of a century (Shukla, Shilpa and Thind, 2000). Addition of quality protein and other functional components by the inclusion of jackfruit seed flour to wheat flour will give a marked improvement in their nutritional value.

Because of increasing awareness in consumers towards health and quality food protein fortification has become a matter of current interest (Tyagi S.K.,2006). The composite flour such as wheat flour blended with soy, corn peanut or mustard flour could be used for the development of bakery items like biscuits and cookies (Tsen, 1976). However, the blended ingredients should not adversely affect the texture, dough consistency, elastic property of the developed dough and undesirable browning reaction during baking (Lorenz, 1983). Available literature on protein fortification with jackfruit seed flour is scanty.

Attempts have been made, in this study, to utilize the potential nutrients of jackfruit seed to develop breads with the utilization of an inexpensive source of protein. It will also add some values with respect to starch and dietary fiber. When developed, this will not only utilize an inexpensive source of protein, but also starch and dietary fiber. The work deals with the nutritional, sensory and textural characteristics of jack fruit seed flour fortified bread tostandardize the extent upto which the jack fruit seed flour may be supplemented in the blend.

(ii) Materials and Methods

Materials*:* Indian Jackfruit seeds (58.43 % moisture content) with white seed coverings were collected from local markets of Kolkata and were used for this study. Commercial wheat flour (100 % machine milled, Ganesh Maida variety) was also purchased from local market.

Methods

***a.* Preparation of jack fruit seed flour:** The seeds (5 kg) were cleaned and the white seed covers were removed. A portion of the cleaned seeds with brown seed coats were then taken for lye-peeling with 2% NaOH for 5 minutes under hot condition (~80^{0}C) to remove the thin brown covering. After washing, the fleshy white cotyledons were collected by scrubbing. Seeds were sliced into thin chips by rotary slicer and then tray dried at (~ 60^{0}C) with a view not to change the functional property much. The time taken was about 16 hours to reduce the moisture content to a safe level of about 10%. The chips were ground to make flour by using a grinder (Bajaj,India). The flour was collected by passing through a 100-mesh sieve and packed in plastic pouch. It was stored at around 8^{0}C in a refrigerator till further use. The yield of the flour was 46%. Proximate composition of the prepared flour was done following the standard methods of AOAC,1984.

b. Flour Blends: To study the effect of blending with wheat flour for the production of bread, the jackfruit seed flour (lye peeled variety) was mixed in different proportions with wheat flour to have 10%, 15%, 20% and 30 % blends for making bread. All the flour samples were packed in polyethylene bags (0.7 mm thickness) and stored.

c. Preparation of Bread: Dry mixing of all ingredients ([flour, sugar, salt, milk powder, GMS (glycerol monostearate), calcium propionate was done. Activated yeast (3% for flour basis) was added. Activation was carried out using lukewarm water and sugar. Melted fat (vanaspati) and then water were added. Dough was developed by proper mixing (mixing time 10 -15 min, till the dough matured).It was kept for 1hour in humid condition for fermentation (at 90% relative humidity condition). Proofing was done at 35-40^{0}C for 1 hour. It was baked in oven for 22-25 min and at 210-220^{0}C.Final bread was taken out and was allowed for cooling. Sensory and chemical analysis were done.

d. Preparation of pan bread: A modified straight dough method (Ranhotra, *et al.*,1974) was used. The base was 350 g wheat flour to which different blends jackfruit seed flour (white flour) in varying proportions was added and the composite flours were mixed separately in a blender , to develop the doughs. A straight dough method, with 1 hour fermentation, 1 hour proofing at 35-40^{0}C and 25 minute baking at 220^{0}C was followed.

Loaves were allowed to cool at room temperature and then placed in plastic bags for sensory evaluation on the second day.

e. Sensory Analysis

Evaluation of bread characteristics: Baking qualities in relation to bread characteristics were evaluated by measuring the loaf volume, the oven spring and the organoleptic scores. Loaf volume was measured by using the rapeseed displacement method, Giami *et al* (2004). Oven spring was determined from the difference in height of dough before and after baking.

Sensory evaluation was performed 24 hours after baking to evaluate loaf appearance, crust colour, crumb colour, taste/flavour and overall acceptability of the bread sample. The bread samples were sliced into pieces of uniform thickness and served with water. Panel members (familiar with quality attributes of local bread) were randomly selected from students and staff of the Department of Food Technology, Techno India to perform the evaluation. Panelists evaluated bread samples on a 9-point hedonic scale as follows: 9 = liked extremely,8 = liked very much,7 = liked,6 = liked mildly, 5 = neither liked nor disliked, 4 = disliked mildly, 3 = disliked, 2 = disliked very much and 1 = disliked extremely (Ranganna,1994)

f. Nutritional analysis*:* The proximate composition such as moisture, ash, crude fiber, crude fat, protein and digestible carbohydrate were determined following the method described by Rangana, 1994 for the jackfruit seed fortified product.

g. Instrumental Measurement: Texture Analysis and Color Measurement

Evaluation of texture Characteristics: The texture characteristics of the bread, were tested using a Texture Analyser (Instron 4301) at room temperature. The samples were analysed with respect to penetration and sponginess. Some modification of the method of Singh *et al*, 1993 have been done setting the instrument probe with a test speed value followed by post test speed. On running the test, the probe penetrated to a certain % of the bread height and then return to its original position. Knowing the resultant force-distance curve the absolute peak force was evaluated. The ratio of the area under the curve was considered as the cohesiveness/ hardness of the bread or in other word the sponginess could be evaluated through penetration test.

Measurement of Color Index: Different blended bread varieties were measured for colour measurement using Hunter Lab Colorimeter, color flex, Colour codes were: lightness(L), redness (+a), greenness (-a), yellowness (+ b) and blueness (-b). The L, a, b values of the standard white tile were 94.25, -0.83 and 0.79 respectively where L = 100 is associated with white and L = 0 is associated with black. All the tests were analyzed in triplicate and the mean results are tabulated.

h. Instrumental Measurement of Micro Structure of Prepared Bread: The SEM (Scanning Electron Micrograph) image was done with the baked dried samples. These were mounted on aluminum stubs using double side tape adhesive glue and then coated with gold (SPI Gold Sputter Coater) (Fornal, 1985) before observing by the SEM. SEM images of samples were taken at 750 x and 1,500 x magnification. The surface morphology of the developed product on the basis of acceptability (jackfruit seed flour blended bread) with respect to their starch and protein granules was studied. Microphotographs were taken with a Jeol JSM 5200 scanning electron microscope (SEM).

7.6 Results and Discussion

a. Proximate Analysis

The jackfruit seed flours which were used for blending and product development, were prepared by removing the brown seed coat with the help of processing and comparable with wheat flour. The prepared flour was subsequently dried mainly for increasing the shelf life for its subsequent use for bakery items as the seeds are seasonal one. However, such treatments are expected to change the functional characteristics of the jackfruit seed flours and hence the enrichment by addition will depend on not only to the extent of addition but also changes which are likely to occur by such treatments. However, the acceptability of the products as determined by the sensory evaluation will be the ultimate governing factor and studied thoroughly. Proximate analysis of the flour was done to have an idea about the nutritional status. The results are listed in Table 1.

Table 1: Proximate Analysis of Wheat and Jackfruit Seed Flours

Proximate analysis %	Wheat flour	Jackfruit seed flour removing brown seed coat
Moisture	11.5	10.1
Crude protein	8.9	12.6
Ash	0.63	2.24
Crude fat	1.4	3.37
Crude fibre	0.78	1.47
Total digestible carbohydrate	76.79	70.22

Sensory Evaluation of Jackfruit Seed Fortified Breads: Different blends were prepared to know the acceptability for breads. The blends tried for bread varied from 10 -30% for nutritional enrichment. The sensory analysis reveals that the substitution of wheat flour with jackfruit seed flour to a level of 15% could develop breads with more acceptable qualities. Though the attributes for breads with 15% and 20% blends showed very close proximity regarding the quality aspects, the loaf volume was found more in case of 15% blended bread variety. More protein enrichment with still higher substitution did not satisfy the acceptance parameters. The report on sensory score was illustrated in Table 2a & 2b

Table 2 (a) Sensory on Bread Characteristics

Sample	Ratio of WF : JFSF in preparing blends w.r.to 100 gm flour	Loaf volume cc	Oven spring cm	Crust color	Crumb color	Flavor	Texture	Overall acceptability
Control	100 : 0	476	1.5	brown	golden	good	good	good
10 % blend	90 : 10	400	2	brown	golden	good	comparab le*	comparable*
15 % blend	85 : 15	530	2.2	little bit darker	bright yellow	gummy highly acceptable	compact	acceptable**
20 % blend	80 : 20	480	2.1	deep brown	bright yellow	acceptable**	more open	acceptable
30% blend	70 : 20	388	0.5	dark brown	golden brown	acceptable	not comparab le	less acceptable**

W.F. = Wheat flour, **JFSF** = Jackfruit seed flour*comparable indicates similarity with control
**acceptable is acceptability as that of control.

Table 2 (b): Score Sheet as Per Sensory Evaluation

Sample	Colour	Flavour	Texture	Taste	Overall acceptability
Control	7	6	5	7	7
10 % blend	7	7.5	5	7	6.5
15 % blend	8.5	7.5	5.5	7	7.5
20 % blend	7.5	7	5	5	7
30% blend	4	6	5	5	5

Sensory evaluation in terms of sensory attributes such as colour, flavor, texture and overall acceptability may vary from population to population and for other reasons. However, the instrumental analysis with the help of Texture Analyzer and Colorimeter may serve as a tool to overcome the vagaries of people-based sensory evaluation. Hence an attempt was made to study the effect of incorporation of jackfruit seed flour in breads instrumentally. The results are recorded in Table 3 - Table 4.

Instrumental Evaluation of Texture of Jackfruit Seed Fortified Breads: Rheology is the study of deformation of materials under stress. Rheological studies provide information concerned with elasticity of solids or viscosity of liquids. Food materials being neither wholly liquids nor solids, exhibit viscoelastic properties intermediate between those of soilds and liquids. Rheology of food materials is describable in terms of elastic deformation, viscous flow and plastic flow.

Rheological properties may be translated into sensory and quality characteristics, like mouthfeel, smoothness, graininess, thickness, firmness, hardness, brittleness, crispiness, general handling properties and many more.

Texture of food stuffs is related to their perceived consistency: hard versus soft, crispy or not, smooth versus lumpy, flows or globs, stands or flops. Texture is determined by the response of the food materials to the applied forces. Rheology is the study of stress and strain. Stress is the force (F) applied per unit area (A) of surface. Strain is the fractional change in length L, Strain = $\Delta L / L$

The two peaks in the graphical trace of Figure 1 represents the deformation or strain of the bread against stress (applied force/ load), the first peak represents mostly the resistance offered by the crust and crumb under first compression and its withdrawal whereas the second peak is for the second compression and its withdrawal. The ratio between the area of work during the second compression and that during the first compression represents cohesiveness (Bourne M, 2002). The pattern of the curve shows the elastic behaviour of each type of product (control, 15% and 30% blended breads).In each case it is observed that the first peak height is more than that of the second one. The cohesiveness was more in case of 30% and was less in 15 % blending compared to control bread and presented in Table 3 . The cohesiveness index was almost similar for both control and 30% blend and lower in case of 15% blending- the preference by sensory analysis was for the lower cohesiveness with less blending and was not that much in two other cases.

Firmness or hardness of bread is a major sensory attribute, related closely to the organoleptic assessment of staleness (Bice and Geddes,1949).Generally these attributes were measured by Instron texture analyzer. Research studies revealed that bread firmness is influenced by several factors, including formulations. Increased skimmed milk solids, for example, significantly increased softness of the crumb .The bread quality with respect to different parameters related with rheology is tabulated in Table 3 with the variation of % blending of jackfruit seed flours.

Our observation also confirms that the formulation has an effect on texture compared to control. The value of hardness as obtained by Texture Analyzer though is higher for blending but organoleptically is not reflected. This indicates that this attribute may not have impact on sensory quality.

Table 3: Bread Quality related to Rheology and Texture

Parameter related with rheology	30% blended bread	15% blended bread	Control bread
p1,hardness	6.8 cm	8.7 cm	3.8 cm
Area2/area1, cohesiveness	0.607	0.5155	0.596
Length2/length1 (base axis) Springiness	0.714	0.642	0.75
Gumminess (Hardness X cohesiveness)	4.128	4.485	2.265
Chewiness (Gumminess X springiness)	2.95	2.88	1.70

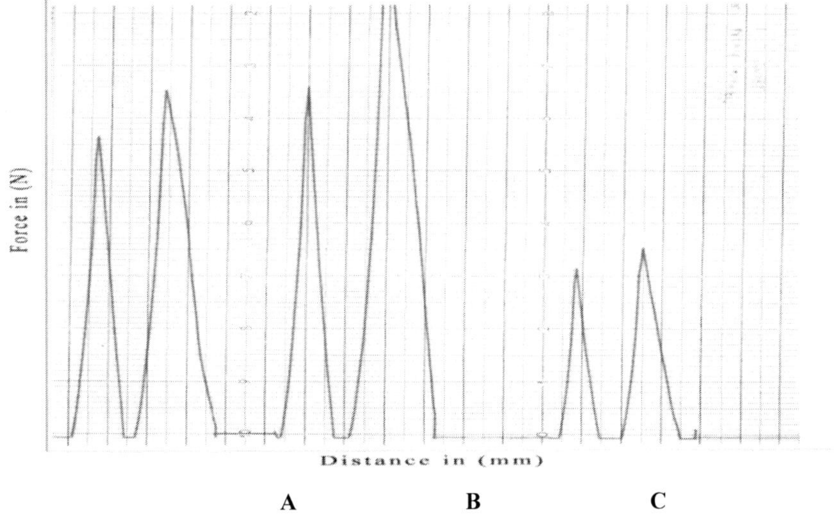

Fig. 1: Textural characteristics of blended breads

In the above graph, the trace of C stands for control bread whereas B and A are for 15% and 30% blended bread variety.

Colour Evaluation using Hunter Lab Colorimeter of Jackfruit Seed Fortified Breads: The results of blending of jackfruit seed flour with the base flour to develop a quality product (bread) was evaluated on the basis of colour and their L, a, b values are presented in Table 4. In case of overall acceptability of bread, colour note plays an important role for crumb and crust both and this could relate to the sensory acceptability of the blended product.

The hue of the product will depend on the values of L, a, b which in turn will be dependent on the extent of substitution. The jackfruit seed flour prepared from the seeds will have a normal tendency to reduce the value of L which will be the index for darkness or lightness. It is also presumed that the substituted flour gets acted upon by the enzymes present in the blended flour releasing more of fermentable sugars providing the conditions for more production of carbon dioxide. Jackfruit seed with high nutritional content is likely to provide some growth factors to the fermenting yeast and the utilization of this fermented carbohydrate will be more by yeast and there by having a tendency to reduce the content of the reducing sugar responsible for browning , though the protein content of the blended flour is high. This is further corroborated by the fact that more of carbon dioxide is produced which is indicated by the oven spring upto 20% substitution. The gluten content of the original wheat flour is reduced by blending that is why the puffiness subsides after certain time because of the weakness in the gluten structure which is also possibly the reason for less oven spring in case of 30% blending. Hence any substitution of wheat flour with jackfruit seed flour will have to take into considerations all these points and can not focus only on protein enrichment. In our studies 15 % blend was found to be the most desirable one considering all factors.

Table 4: Measurement of colours Blended bread by Hunter Lab Colorimeter with the Variation of Flour Blends

Sl.No	Bread sample	L	a	b	Delta E
1.	30% blending	Crumb—49.24	6.66	16	48.097
		Crust—39.60	7.70	13.02	56.65
2.	20% blending	Crumb—64.23	5.17	18.19	35.21
		Crust—47.90	9.57	17.02	50.19
3.	15% blending	Crumb—65.18	5.23	17.95	34.30
		Crust—47.83	9.57	16.74	50.17
4.	10% blending	Crumb—70.23	4.68	17.72	28.87
		Crust—49.67	9.63	17.01	48.58
5.	Control bread	Crumb –92.63	0.81	1.39	2.66
		Crust—30.27	5.83	7.20	64.64

Instrumental Evaluation of Morphology of Jackfruit Seed Fortified Bread and Cookies: SEM pictures (1500 X or 750 X) of jackfruit seed starch granules along with the wheat flour starch upon gelatinization was characterized. The cooking or baking of composite wheat flour, blended with jackfruit seed flour, showed an impact on the product quality. And accepted products on sensory basis depend on the formation of gelatinized matrix of starch and protein fractions. The processing operation develops the product characteristics like texture, flavor and color, primarily by gelatinization of the starchy fractions . Microscopy technique and the images were used to judge the quality of starch gelatinization in baked flour matrix. Small pores were observed in the picture of the gelatinized starch matrix which might be the indication of the presence of protein molecules. Similar observation was reported in maize starch by Juszczak, 2001and Pa³asiñski *et al.*, 2000 who claimed that these regular pores were probably formed during granule growth as an imprint of globular proteins of endosperm.

Microscopy image was used to judge the quality of starch gelatinization in cooked blended flour matrix for bread and presented in Figure 2

Fig. 2: SEM Image of Jackfruit Seed Flour Blended Bread Crumb

Nutritional characteristics of jackfruit seed fortified bread : The proximate analysis was done with the product having 15% blend of jackfruit seed flour because of its highest score on sensory acceptance basis. The protein enrichment was nearly 1% by quality protein in case of blended bread as reported in Table 5. The prepared bread was also slightly rich in crude fibre and mineral content compared with control bread.

Table 5: Nutritional Evaluation of Bread of Best Variety Proximate Evaluations of bread with 15 % blend on dry basis

Parameters	Wheat flour bread	Blended bread
Moisture (%)	39	43
Ash (%)	0.86	1.44
Protein (%)	11.06	12.24
Crude fibre (%)	0.48	1.42

The present work suggests that bread could be developed by blending with 15 % jackfruit seed flour with wheat flour on overall acceptability basis. The use of lye peeled 60^{0}C,100 mesh processed jackfruit seed flour in preparing bread had the familiar color of bread which were almost comparable with the control. The over all sensory study on the bread quality indicated that 15% blended bread with 60^{0}C,100 mesh jackfruit seed flour had proper volume, shape and size. The water retention property remained good for such blended bread which was reflected in retention of the soft and spongy texture over a period of storage.

The effects of substituting 5, 10 and 15% field pea or defatted soy flour for wheat flour in a chemically leavened quick bread on physical characteristics of batters and bread were investigated by Raidl and Klein (1983). They studied the sensory characteristics of bread also. Significant differences were observed for batter spread, consistency, loaf volume, darkness and yellowness value of bread crumbs. The different types of bread prepared with different % blend of jackfruit seed flour with respect to base wheat flour also had some differences in product quality. The varied crumb and crust colour was seen with varied % composition of flour blends which is summarized in Table 4. Results indicated that development of breads with incorporation of jackfruit seed flour at levels upto 15% was successful, contrary to the adverse effect on both physical and sensory characteristics of bread during incorporation of field pea flour. Volumes of loaves, studied by Khan *et al.* (1975), with peanut protein concentrates at levels higher than 10%, were significantly lower than those of loaves made from flour containing other protein supplements. The loaf volume was found to be highest as 530 cc with bread having 15% of blending of jackfruit seed flour with wheat flour among all types of blending and reported in Table 2a.

Conclusion: The present study reveals that jackfruit seed flour can be used as a potential supplement in new food formulation along with wheat flour. The 15% (w/w) blending of jackfruit seed flour with wheat flour could be used in developing bread with comparable sensory acceptability with control. The flour blends could be used as protein supplements and functional ingredients in human diets. Chemical analysis reveals that the substituted breads are protein and fibre enriched. The proximate analysis of the bread suggests that protein enrichment could be achieved with jackfruit seed flour blend (15%) upto 12.24% (on dry basis) where as protein content of the control was 11.06%. The proper volume , shape and size, softness and sponginess were found to be maintained on storage for 3 days. Thus consumption of bread supplemented with jack fruit seed flour could be beneficial for improving nutritional status in India with respect to protein and crude fibre content. This attempt in developing such bakery items could address the scarcity problem of non-wheat producing countries and even during scarcity of wheat production or unavailability of wheat flour in local market.

Case Study - II

7.7 Spices Sector

A. Waste Generation from Spice-processing

Spices and herbs are presently the most current focus in the field of agro-commodities. Indulgent aroma and taste of food come from seeds, leaves, stem and other parts of plant. They are responsible for specific therapeutic properties such as antioxidative, anti-inflammatory, antidiabetic, antihypertensive and antimicrobial activities (Samah,2019)

The two major processed products from spice and herb produce in both domestic and export market are essential oil and oleoresin. Extraction of these two fractions generate spice/herb spent residue, which is almost 80-90% of the raw produce. Spent residue, treated as waste from spice-processing, till now, is not utilized to a great extent in food and feed industry. Increase in export of spices from 7265 ton (2011-12) to 11635 ton (2015-2016)also increases amount of spice-spent generation and thus creating disposability problem with effect on environmental pollution. Transformation of this residue to other consumable products would definitely result in a sustainable solution for spice and herb- processing industry (Sowbhagya,2019).

B. Value Added Components from Spice Wastes

There is a strong need for development of value-added products. Ground spices have a limited shelf- life and the flavor loss or alteration of flavor happens on storage. Dispersed spices spread on dextrose, salt, husk, etc. tend to lose the flavor on storage due to significant surface area. Dehydration is also not a suitable technology to retain its natural flavor. Spice-oil and oleoresins have their own limitations regarding unbalanced and concentrated flavor-profile respectively and uniform dispersion in the final product. Increasing demand for convenience foods require customized and stable flavor- profile, more suitably achieved by blending of synthetic and natural flavor(www.jnkv.

org.). Currently, spice-processing industry generates residue, which becomes a waste without any commercial applications. But this can be an effective food ingredient, as it is a rich source of functional phytocomponents. Utilization of this industrial waste effectively curb environmental pollution from residues of spice-processing industries (Sowbhagya *et al.*, 2007).

7.8 Research-Based Study

Value-addition of Spice-processing By-product Through Physical Modification

Introduction

Spice spents, obtained as residue from essential oil and oleoresin processing plant, are found to be enriched in dietary fibre, protein, vitamins, polyphenols and vital minerals like calcium, iron, magnesium and zinc. Dietary fibre content of more than 50% has made this residue an important functional ingredient to be used in bakery products (biscuit, bread), low in fibre content. Literatures have been already found for incorporation of spice residue of many spices such as chilli, cumin, celery, ginger, turmeric etc.

Another approach for value-addition of spice-residue would be to extract and isolate functional ingredients from it, such as starch and other bioactive components. Santana *et al.*, (2019) generated turmeric solid waste after application of supercritical fluid extraction for deflavouring and pressurised liquid ethanol extraction for extraction of curcuminoids. A mixed biopolymer having relevant composition of starch along with antioxidant constituents was isolated from solid turmeric waste (Santana *et al.*, 2017).

Solid spice waste has been exploited much to recover starch. Solid turmeric was macerated and blended in mixer-blender maintain a ratio of turmeric and water of 1:6. The macerated mass was then filtered through nylon mesh and the filtrate was allowed to centrifuge. The filtration pie was resuspended in water and inserted into centrifuge at 1700–1800 g for 10 min .Isolated starch was again washed and suspended in ethanol and was filtered through vacuum pump. It was then dried at 40-50C for 4 hr to get turmeric starch (Santana *et al.*, 2017) .On dry basis, percentage yield was reported as 3.33, which is supported by previous investigations obtained for arrowroot starch, 7.07- 8.50% (Pakira *et al.*, 2014).

Earlier investigators reported significant hot-peak viscosity (1.29 mPas) for turmeric starch paste containing 10% solids. (Dhanalakshmi *et al.*, 2011). Santana *et al.* also characterized the products obtained from turmeric waste and found 0.3-13 g. curcuminoids/100g of product and starch was quantified from 26-46 g/100g. These properties indicate potential applications of these products in foods and other personal care products.

Maniglia *et al.* reported mechanical and chemical treatment on turmeric starch obtained from turmeric solid waste using alkali (NaOH), sodium hypochlorite(NaOCl) and hydrogen peroxide. Structural modification was resulted by this treatment to form films. Literature related to physical modification of turmeric starch is scanty. Therefore, the present work aimed at physical modification of turmeric starch to investigate the functional properties of modified starch.

Materials and Methods

i. Materials: Native turmcric starch, distilled water ii. Methods: Modification of native turmeric starch:

a. Physical modification, heat-moisture treatment, was carried out on native turmeric starch according to the method followed by Korrakot (2006). Native starch powder was spread in petridish. Water was spread to bring the moisture content up to 25%. Moist turmeric starch was placed in plastic bag & sealed. It was kept for 24 hours at 2°-4° C temperature. Then starch sample was divided in two parts in two Petri-plate. Starch sample was placed in a hot air oven at 120° C for 1 hour & 3 hours. Next, starch sample was dried in tray drier at 45-50° C temperature. Then it was stored in a zip- lock LDPE pouch.
b. Physicochemical analysis of native & modified turmeric starch:

Native turmeric starch alongwith heat-moisture treated starch (1 hr. and 3 hrs.) were analyzed for determination of swelling power and % solubility.

Determination of Swelling Power

According to Wattanachant *et al.* (2002), swelling power and percentage solubility of starch-water suspension were determined. Briefly, weighed starch sample was suspended in distilled water in a 250 ml. beaker through constant stirring process. It produced a uniform dispersion. Suspension was then heated in a thermostatically controlled water-bath, set at temperature 85C with constant agitation for 30 minutes. It was then centrifuged for 15 minutes at 5000 rpm. Swollen mass was obtained as residue. Soluble fraction of starches are recovered from supernatant by evaporation of the liquid. The amount of this material was used to calculate percentage starch solubility. Swelling power was obtained by measuring weight of residue from centrifugation considering content of water absorbed bystarch (percent weight increase).The result was calculated by equation (1) (Wattanachant, *et al.* 2002), considering sample weight after due correction for the amount of solubilized starch.

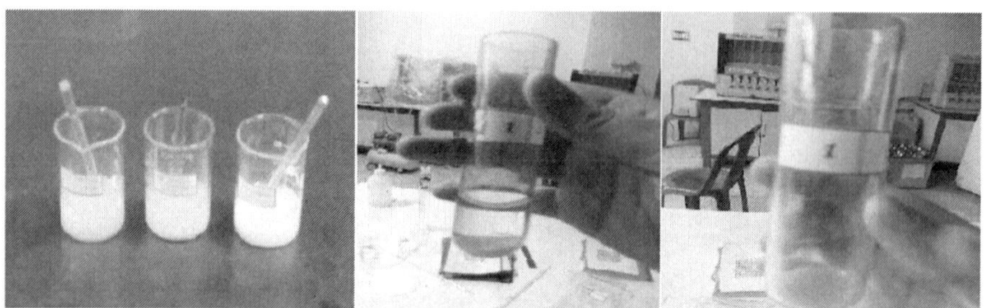

Swelling power = Weight of sediment paste x 100/Weight of sample on dry basis x (100 -%solubility) (1)

7.9 Results and Discussion

Table 6: Comparison of swelling power and solubility between native and modified starches:

Sample details	Native starch	HMT 1 hr.	HMT 3 hr.
Sample weight	1.0302	1.0063	1.0744
Sediment paste	4.2937	4.3431	4.2937
Volume of supernatant	19	20	21
Moisture content	9.16	9.25	10.1
Residue	0.0555	0.0718	0.0744
Total residue in supernatan	0.1054	0.1436	0.1562
% solubility	10.24	14.27	14.52
(100-% solubility)	89.76	85.73	85.48
Swelling power	4.643	5.034	4.675

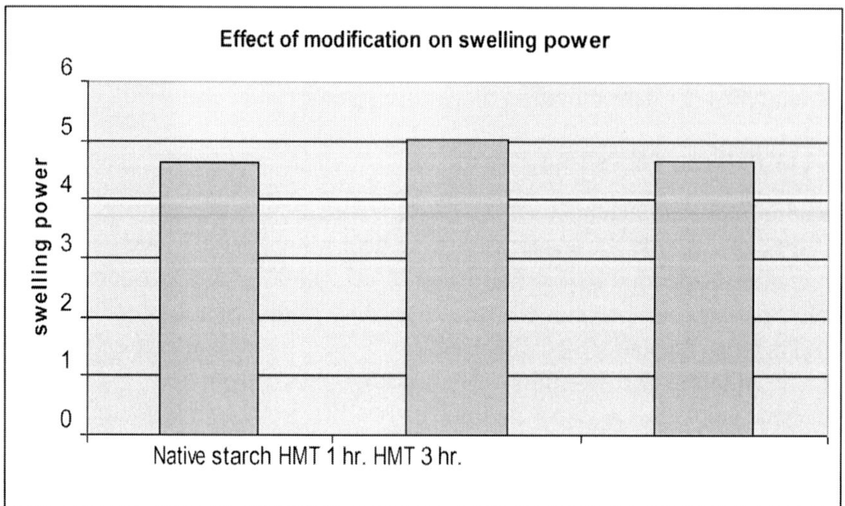

Swelling property basically indicates extent of water absorption by starch granules and its susceptibility towards heating in an aqueous solution.

Both swelling power and solubility tend to increase in value as temperature increases, but this trend also depends on type of starch granules significantly.

Swelling power behaviour of native and modified turmeric starches was shown by this graph. There was not much change in swelling power between native and modified turmeric starch. Swelling power range for these starches was found to be 4.64-5.03(g/g).Earlier findings report swelling power of ginger starch as 3 (g/g) (Bello *et al.,* 2013).

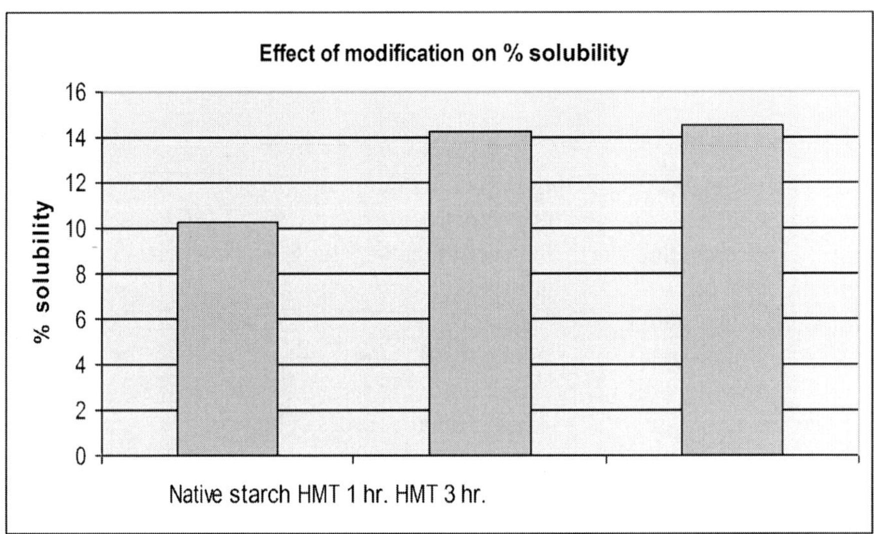

This graph represents effect of modification on % solubility of native and modified starches.It has been observed that % solubility increased much (1.4 times) after modification. Results showed that the % solubility value of native turmeric starch (10.24) was much higher than earlier reported value of ginger starch (Bello *et al.*, 2013). After modification intermolecular bonding of starch granules might get weakened, which in turn increased leaching of amylose fraction and percentage solubility of native turmeric starch.

Conclusion

Heat-moisture treatment resulted in change in swelling power and % solubility of native starch. Variation in process time also showed difference in physicochemical properties. Increase in % solubility of heat-moisture treated starch indicated that structure of native starch was altered. This facilitates its use as a thickener in soup-premix or in fruit or vegetable-based beverages.

7.10 Applications of Products Recovered from Turmeric Waste

Turmeric starch can be used as a potential source of non-conventional starch. The turmeric-spent, rich in starch fraction, also showed potential to be used for preparation of biofilms for packaging. Temperature and pH were optimized to produce biodegradable film with 9MPa mechanical strength, 37% solubility and 0.352 gm/hr/m^2 water vapour permeability (Maniglia, 2015).Turmeric spent flour, rich in starch and antioxidants, was blended with raw rice and black gramat different proportion to develop Indian pan-fried snack, dosa. Colorimetric analysis and sensory analysis showed comparable value with the control.

7.11 Conclusion

Fruit and vegetable processing units include juice processing, pulp extraction, making jams jellies marmalade and frozen pulp, canned products and many more. These generate waste and such wastes include skin, pulp, seed waste and pomace, which are rich in value added compounds like starch, fibres, proteins etc. Increasing demand for specialty products like starch or protein with desirable functional or physicochemical properties and demand for supply of food for growing world population motivate search for an alternative source of protein or starch other than conventional sources .The principal conventional sources of starch are cereals and tubers whereas proteins from vegan sources are pulses and lentils. However, fruit industry and spice processing waste can be recycled with respect to isolation of significant amounts of starch and protein and it can be better used back in food matrix.

7.12 References

Agrawal SR 1990. Prospects of small scale biscuit industry in the nineties, Indian Food Industry, 9, 19.

Ahmed H and Chatterjee BP 1989. Further characterization and immunochemical studies on the carbohydrate specificity of jackfruit (*Artocarpus integrifolia*) lectin. J. biol. Chem., 264, 9365.

Ajila CM, Bhat SG, Rao UP. 2007. Valuable components of raw and ripe peels from two Indian mango varieties. Food Chem 102: 1006– 11.

Ayala Zavala JF, Rosas Domínguez C, Vega Vega V, González Aguilar GA. 2020. Antioxidant enrichment and antimicrobial protection of fresh cut fruits using their own byproducts: looking for integral exploitation. J Food Sci, 75, 175– 81,

Bice CW and Geddes WF 1949. Studies on bread staling IV. Evaluation of methods for measurement of changes which occur during bread staling. Cereal Chem., 26, 440

Bobbio FO, El-Dash AA, Bobbio PA & Rodriguis LR 1978. Isolation and characterization of the physico-chemical properties of the starch of Jackfruit seeds (Artocarpus heterophyllus),Cereal Chem, (55), 505

Bourne M 2002. Food texture and viscosity: Concept and measurement. Academic Press, 353.

Burkill HM 1997. The Useful Plants of West Tropical Africa,.Vol. 4, 2nd Edn., (Royal Botanic Gardens, Kew), 160.

Buzby,J. Hyman, J., Stewart, H. , Wells, H. 2011. Value of Retail- and Consumer- level Fruit and Vegetable losses in the United States – Journal of Consumer Affairs 45(3) : 492-515

Campbell RJ, El-Sawa SF & Eck R 1998., The jackfruit, Fairchild horticulture series , Vol. 2. Fairchild Tropical Garden, Miami, FL, pp 23.

Dhanalakshmi, K., Rao, L.J.M. and Bhattacharya ,S. 2011.Turmeric Powder and Starch: Selected Physical, Physicochemical, and Microstructural Properties. Journal of Food Science.November. 1284-91

FAO. 2014. Definitional framework of food losses and waste. Rome, Italy: FAO.

Favero J, Corbeau P, Nicolas M, Benkirane M, Trave G, Dixon JF, Aucouturier P,Rasheed S, Parker JW, Liautard JP. Inhibition of human immunodeficiency virus infection by the lectin jacalin and by a derived peptide showing a sequence similarity with gp120. Eur JImmunol ,,23, 179,1993.

FO Bobbio, AA El-Dash, PA Bobbio and LR Rodrigues, I1978. solation and characterization of the physicochemical properties of the starch of jackfruit seeds (*Artocarpus heterophyllus*), Cereal Chem, 55, 505-11.

Fornal J 1985. Influence of hydrothermal treatment on the starch of corn grain. Acta Alim. Pol.,11, 141.

Garnett T. Fruit and vegetables and UK greenhouse emissions: exploring the relationship. UK: Food and Climate Research Network, Univ. of Surrey.,2006

Giami SY, Amasisi T, Ekiyor G. 2004. Comparison of bread making properties of composite flour from kernels of roasted and boiled African breadfruit (*Treculia Africana decne*) seeds. J. Raw Mat. Res.,1, 16

Gustavsson J, Cederber C, Sonessan U, Otterdijk R, Mcybeck A. 2001. Global food losses and food waste: extent, causes and prevention, Food and Agricultural Org. of the United Nations.

HLPE 2014. Food losses and waste in the context of sustainable food systems.A report by the high level panel of experts on Food Security and Nutrition of the committee on World Food Security, Rome.

http://www.jnkvv.org/PDF/05042020135315spices.pdf

https://usgreentechnology.com/green-technology/

Joshi VK, Kumar A, Kumar V. 2012. Antimicrobial, antioxidant and phyto-chemicals from fruit and vegetable wastes: a review. Intl Food Ferm Technol, 2, 123– 36, 2012.

Juszczak L 2001. Surface structure of starch granules , Zeszyty Naukowe AR w Krakowie , 13, 73 (in Polish).

Kader, A.A. 2005. Increasing food availability by reducing postharvest losses of fresh produce, Acta Horticulturae, vol.682

Khan MN, Rhee KC, Rooney LW and Cater CM (1975), Bread baking properties of aqueous processed peanut protein concentrates. J. Food Sci.,40, 580.

Kolawole SA, Igwemar NC, Bello HA.2013. Comparison of the Physicochemical Properties of Starch from Ginger (Zingiber officinale) and Maize (Zea mays).ijsr,2,11

Korrakot, L and Onanong, N. 2006. Modification of Rice Flour by Heat Moisture Treatment (HMT) to Produce Rice Noodles; Kasetsart .J. (Nat. Sci.), 40,135 - 143

Kumar S, Singh AB, Abidi AB, Upadhyay RG & Singh A 1988. Proximate composition of Jackfruit Seed, Journal of Food Sci and Technol, (25 , 308-309.

Kummu, M. Moel, H. Porkka, M. Siebert, S., Olli, V., Wara, P. 2012. Lost food, wasted resources: Food supply chain losses and their impacts on freshwater, cropland and fertilizer use. The Science of the total environment. 438C

Lionel, M., Sarmanto, S.D.S, Cereda, M.P. 2003. New starches for the food industry: curcuma longa and Curcuma zedoaria, Carbohyd. Polym., 54,3,385-388.

Lorenz K 1983., Protein fortification of cookies. Cereal Food World, 28(8), 449.

Mahanta SK, Sastry MVK and Suroha A 1990. Topography of the combining region of a Thomsen- Friedenreich-antigen-specific lectin jacalin (*Artornrpus in - tegrifoliu agglutinin*). Biochem J. , 265, 831.

Maniglia, B. C., de Paula, R. L., Domingos, J. R., Tapia-Blacido, D. R. 2015. Turmeric dye extraction residue for use in bioactive film production: optimization of turmeric film plasticized with glycerol. LWT – Food Science and Technology, 64: 1187–1195.

Othman S, Joudu I, Bhat R. 2020. Bioactives from Agri-Food Wastes: Present Insights and Future Challenges. Molecules, 25, 510 .

Pa³asiñski M, Fortuna T, Juszczak L, Fornal J 2000. Changes in some physico-chemical properties of starch granules induced by heating and microwave radiation, Pol. J. Food Nutr. Sci., 6/50, 17.

Pakira, C.K., Patel,S;Naik, RK, Ram, D.S. 2014. Performance evaluation of tikhur (*curcuma angustifolia*) starch extraction machine. Ind. J. Sci. Res. Tech., 2,5,34-38

Rahman MA, Nahar N, Mian AJ & Mosihuzzaman M 1999. Variation of carbohydrate composition of two forms of fruit from jack tree (*Artocarpus heterophyllus* L) with maturity and climatic conditions, Food Chem, 65, 91.

Raidl MA and Klein BP (1983), Effect of soy or field pea flour substitution on physical and sensory characteristics of chemically leavened quick breads, Cereal Chem., 60, 367.

Ranganna S 1994. Handbook of analysis and quality control for fruit and vegetable products. New Delhi : Tata Mc Graw-Hill Publishing Co.,Ltd.,

Ranhotra GS and LoeweR.J (1974), Bread making characteristic of wheat flour fortified with various commercial soy protein products. Cereal Chemistry,51, 629,1974.

Sagar N, Pareek S, Yahia E ,Lobo G. Fruit and 2017. Vegetable Waste: Bioactive Compounds, Their Extraction, and Possible Utilization. Comprehensive Reviews in Food Science and Food Safety, 17,3.

Sajitha, P.K. and Shashikumar, B. 2015. Qualitative and quantitative variation in starch from four species of curcuma Cytologia, 80 ,1, 45-50.

Samah M, El-Sayeed, Ahmeed L 2019. Youseef. Potential application of herbs and spices and their effects in functional dairy products. Heliyon; Vol 5, issue 6.

Santana AL, Meireles MA. 2014. New Starches are the Trend for Industry Applications: A Review. Food and Public Health., 4,5.229-241

Santana, AL, Giovani,L, Tobon, FO, Joshner, JCF. 2018. Partial hydrothermal hydrolysis is an effective way to recover bioactives from turmeric wastes. Food Sci. Technol, Campinas, 38(2): 280-292, Apr.-June.

Santana, AL, Giovani,L, Tobon, FO, Joshner,JCF. 2017. Starch recovery from turmeric wastes using supercritical technology. Journal of Food Engineering,214, 266-276

Schieber A, Stintzing FC, Carle R. 2001. By products of plant food processing as a source of functional compounds recent developments. Trends Food Sci Technol 12: 401– 13.

Shukla FC, Shilpa M & Thind SS. 2000. Bakery industry in India – present quality control and future scenario - A review. Beverage and food world,27, 11

Singh B, Bajaj M, Kaur A, Sharma S & Sidhu JS 1993. Studies on the development of high protein biscuits from composite flours. Plant Foods for Human Nutrition , 43 (2), 181.

Sowbhagya, HB, Suma, PF , Mahadevamma, S. and Tharanathan, RN 2007. Spent residue from cumin - a potential source of dietary fiber. Food Chemistry, 104 (3), 1220-1225,2007

Sowbhagya,HB. 2019. Value-added processing of by-products from spice industry. Food Quality and Safety, 3,73–80

Tanuja S, Chatterjee U, June H. Wu, Bishnu P, Albert M. Wu. 2005. Carbohydrate recognition factors of a Tα (Galβ1→3GalNAcα1→Ser/Thr) and Tn (GalNAcα1→Ser/Thr) specific lectin isolated from the seeds of Artocarpus lakoocha. Glycobiology,15, 1, 67-78.

Tsen CC. 1976. Regular production of protein fortified cookies from composite flours. Cereal Food World, 21 (12), 633.

Tyagi SK, Manikantan MR, Oberoi HS & Kaur G. 2006. Effect of mustard flour incorporation on nutritional , textural and organoleptic characteristics of biscuits.Journal of Food Engg,,80, 1043.

Valorization of Food Processing By-Products. Ed. M. Chandrasekaran, CRC Press, Taylor & Francis Group 2013.

Vedashree M, Pradeep K, Ravi R and Madhava Naidu M. 2016. Turmeric Spent Flour: Value Addition to Breakfast Food. Int J Nutr Sci,.1,2, 1006.

Wattanachant S, Muhammad SKS, Mat Hashim D and Rahman RA. 2002. Suitability of sago starch as a base for dual-modification. Songklanakarin J. Sci. Technol.,24,431-438.

8

Various Microencapsulation Process Technologies Used in Food Industries An Overview

Sumit Sudhir Pathak* **and** ***Mahipal Singh Tomar***

ABSTRACT

Microencapsulation is one of the latest technologies used in the food, pharmaceutical, nutraceutical and cosmetic industry where it is used for maintaining the stability of the food ingredients and slow release of the active food ingredients. For many decades the encapsulation technique was used in the pharmaceutical industries, but from last three decades it has been effectively used in the food industries showing enhanced results in flavour masking and retention of bio- active components present in the food. The production of the microcapsules can be achieved by the various process technologies. The process technologies have been reviewed in this paper such as spray drying, extrusion, spray chilling, emulsions, fluidized bed coating, freeze drying, molecular inclusion, crystallization, co-acervation etc. the encapsulation of food materials has shown a promising result in the newly growing food processing industries. Some of the recent developments in the microencapsulation viz. sweetener immobilization, controlled release and the carrier materials are also being presented in the reviewed paper.

Keywords: microencapsulation; microcapsules; food ingredients; extrusion; freeze drying; flavour masking.

8.1 Introduction

The entrapment of a pure material into another carrier material is known as encapsulation. By using this technology the stability of material can be enhanced as well as it can be used to maintain the viability. The active ingredient can be protected from the moisture, external heat and other extreme conditions. The food industry have shown interest in many encapsulation of active ingredients such as essential oils, anti-oxidants, artificial sweeteners, vitamins, minerals,

*Corresponding Author

enzymes, amino acids, probiotic micro-organisms, lipids, acids, bases and buffers, flavoring agents, pigments and dyes. The encapsulation technique was earlier restricted as it was an expensive technique requiring skilled persons to handle and execute the process. But since the volume of production have increased. The scientists, researchers and industrial people have known the fact regarding the techniques specificity and accuracy, the encapsulation technique have gained popularity. Now a days, due to the higher volume of production the cost has been lowered. There are number of encapsulated foods ranging in the nutrition products are gaining popularity among the masses. As the technique of encapsulation have reduced the nutritional losses of food which were occurring during the processing. The nutritional stability have attained by various products were oxidation prone minerals and vitamins were present. Such all types of problems are removed by the usage of encapsulation technique. On very broader sense, we can say that the very minute or fine particles of ingredients having a nutraceutical value or a special function such as emulsification, bio-active components, anti-oxidants, sweeteners etc. are being coated by secondary materials which would provide a protection layer to the active ingredient particle encapsulated inside.

When tiny particles quadruplets are surrounded by a coating which are embedded inhomogeneous or heterogeneous matrix which will be used as a small all capsule having multiple useful properties. This process of production of the microcapsules is called as microencapsulation. When the food ingredients, cells, enzymes and other materials are incorporated in the small capsules is called as encapsulation which can be achieved by the various process technologies as physical, chemical and physicochemical properties. For the capsule formation or shell formation, dextrin, alginates, fats, starch are used in food industry(Reineccius & Bangs,1985).

The particles which are prepared for the formulation of the microcapsule depends on the physio- chemical properties of the composition of the wall materials and core materials. Particles which can be obtained are of possessing uniform thickness coating on simple sphere, core with an irregular shape, a continuous matrix possessing many core particles, multiple cores in same capsule, microcapsules containing multi-wall (Ono & Aoyama, 1979).

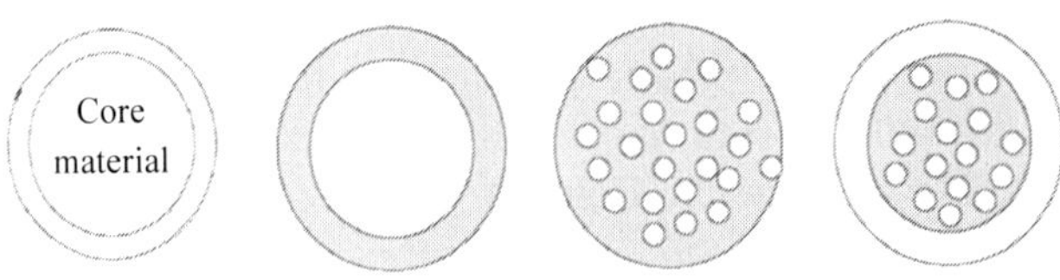

A. Core material B. Round Matrix C. Core Coated in round Matrix

Fig. 1: Structure of core and matrix (*Source*: Fang, & Bhandari, 2010)

Chemical functionality, polarity, solubility and volatility governs the retention of cores. These cores are made up of single or multiple ingredients possessing the single wall or multiple layer wall. A uniform wall with a small sphere is the simplest form of microcapsule, it may have multiple wall. The internal phase or core is always inside the microcapsule while the external phase or wall is called as coating material or shell (Flinn&Nack, 1967). A crystalline material, an emulsion, an adsorbent can be used as core. The figure 1 shows the structure for the core and matrix material. While the figure 2 is showing the formation of microcapsule and microsphere (Kondo, 1980).

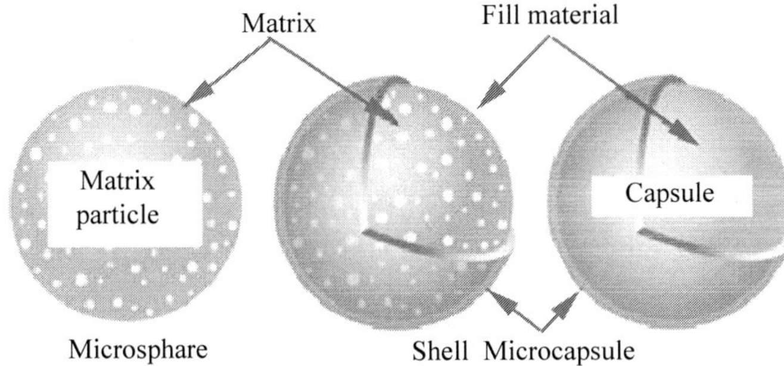

Fig. 2: Structure of microsphere and microcapsule
Source: Oxley, 2014

The structural design of micro capsule can be of following shapes

1. Irregular shaped
2. Core shell microcapsule
3. Polynuclear micro capsule
4. Insoluble matrix microsphere
5. Soluble matrix microsphere

The two key applications served by the microencapsulation technology in the food industry are as follows:

The active ingredients present in the food products can be provided by a good stability by using microencapsulation. The microencapsulation offers and improved performance to food products having active ingredients which includes coffee, chewing gums, probiotics health foods, candies, minerals, vitamins, and enzymes (ONO, 1980). A desired physio-chemical reform in the sensory analysis and perception of the food commodity at an appropriate time or by utilizing a suitable mechanism can be brought by the primary application of Microencapsulation Technology (Martín Banderas et. al., 2005).

The advantages of Microencapsulation are enlisted as below:

1. Orders for taste will be masked., The core taste will be masked.
2. Stability will be enhanced.
3. Viability will be maintained.
4. Moisture content will be protected against heat and extreme conditions.
5. Lumping will be prevented.
6. Easier handling of core material.
7. Evaporation for transfer rate will be decreased.
8. Reactivity of the core will be reduced as the interaction with light, water and oxygen will be restricted.
9. Conversion of liquid form to solid form.
10. The right stimulus is achieved by controlling the release of core material.
11. Uniform dispersion can be achieved in the host material.

The desired morphology, stability and release mechanism can be provided by selecting the appropriate process (Sparks, 1989). While selecting it, one must also refer to the economic feasibility for the industrial scale production, capital investment cost, operational cost and other regulatory expenditures including the storage and transportation and handling. Few of this techniques are being discussed in the review.

8.2 Technologies for Microencapsulation

(i) Spray drying

One of the oldest commercial techniques is spray drying for the production of encapsulated flavors. It was found that acetone added to the tomato puree while producing the spray dried tomato powder, it retained the color and flavor compounds present in the tomatoes. Since 1930, spray drying has become most popular and reliable method in food industry for encapsulating the flavors.

For the preparation of microencapsulation by spray drying technique the component is dissolved in an aqueous solution. The dispersed phase of the solution is homogenized. The particles are atomized and the dehydration of particles takes place(Guzey&McClements, 2006). The particle size is of 10 μm to 400 μm. It is one of the simplest and the fastest way to scale up. The production cost of microcapsules produced by spray drying method is 32 to 50 times cheaper than any other method used for production of microcapsules. In thus technique both the types of polymer are hydrophobic and hydrophilic can be utilized (Goud and Park, 2005). The spray coating technique is shown in the figure 3.

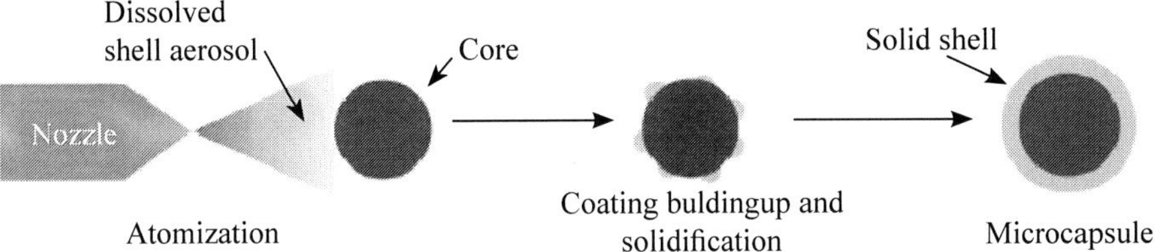

Fig. 3: Formation of microcapsule by spray drying
Source: Abbas *et al.*, 2012

A suitable carrier material has to be selected before performing any encapsulation. The carrier material should possess many properties such as low viscosity, low hydroscopicity, bland taste, and good film forming nature, flavour releasing ability when reconstituted as well as low in cost. The carrier material in which active or flavour components are encapsulated should protect them from heat to avoid the flavour losses. Gum Arabica and hydrolysed starches are the ideal carriers used in food industry which satisfy all the above mentioned properties.

(ii) Spray Cooling or Spray Chilling

Spray chilling is another method which is used for encapsulating the minerals, vitamins and acidulants. The outer membrane forming substance used in the spray chilling is generally a fractionated or hydrogenated vegetable oil. The material which is encapsulated is mixed with these carriers and atomized at a low temperature depending on the active compound to be encapsulated (Risch 1995; Taylor 1983). The materials which are heat sensitive and are insoluble in regular solvents can be encapsulated by using this technique. Dry soup re-mixes; specialty bakery products or food items with high fat content are the commercial examples for spray chilling (Blenford 1986).

(iii) Fluidized Bed Coating

When a solid particle gets deposition of atomized droplets is known as spray coating which can be practiced in the many forms under the names such as rear suspension coating Pan coating fluidized bed coating and granulation (Brown, *et al.*, 2003).

In the Fluid bed coating the core particle are fluidized and added when there is a coating solution; the coating of the core particles is dehydrated cool down to get the core and shell structure (Gouin, 2004). We will always get uniform layer of the shell material over the core solid particle. The fluidized bed coating technique can be applied to all shell materials such as proteins, fats, emulsifiers, yeast, polysaccharides, cell extracts etc. The melted fats, emulsions and waxes are new molecules used as coating materials in the encapsulations but having very promising results. The fluidized bed coating has many advantages over other older techniques of encapsulation. One of the major advantages is of controlled release of encapsulated active component. While the disadvantage of Fluid bed coating is that the air temperature and airstream controlling is a very critical factor. It is also difficult to achieve the uniform coating over every droplet which are significantly smaller than the core (Garti, & McClements, 2012).

When the wet droplets are accumulated around the solid core particles they will be solidified by the drying process. by the subsequent cognolin of such wet droplets around the dry surface which will result it in the formation of Shell material which is known by solid Core.

(iv) Co-extrusion

Co-extrusion is referred as an alternative form for the annular jet atomization. Concentric nozzle system is used for the formation of core shell droplets. The breakup of an annular jet through the axisymmetric and forms the outer shell of the droplet which will solidify to form a micro capsule shell. The exclusion can be undertaken by various techniques such as submerged nozzle extrusion stationery co extrusion centrifugal extrusion and vibrating nozzle extrusion(Desai, & Jin Park, 2005). In the exclusion produced microcapsule the product has a very long shelf life of 5 years. only disadvantage of using the exclusion Technology is large particles are formed during extrusion while it be very limited shell materials are available for the extrusion(Gibbs et. al., 1999). The figure 4 is showing the microsphere formation by using the extrusion.

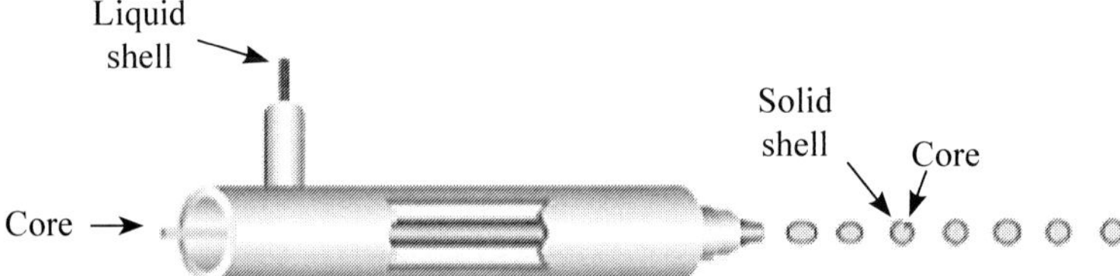

Fig. 4: Formation of microsphere by extrusion
Source: Anjani et. al., 2007

(v) Emulsion

In the first stage of the microsphere formation, the emulsion is to be prepared, and in the second stage a core material is dispersed in system possessing an immiscible liquid phase. Here, shell formation takes place around the droplets in dispersed phase (Frascareli *et al.*, 2012). The oil in water is very common example which is used for the encapsulation. The figure 5 is showing the formation of microsphere formation by using the emulsion.

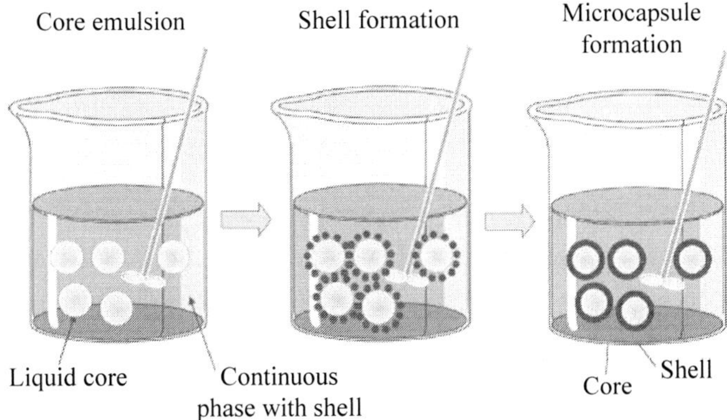

Fig. 5: Formation of microsphere by emulsion
Source:Oxley, 2014

(vi) Freeze Drying or Lyophilization

Freeze drying is applied to produce microcapsules. The core is mixed in the coating solution, then complete mixture is freeze dried and finally grinded to get the Powder. The product obtained from the freeze drying has a very good resistance to oxidation it can also maintain the structure of the micro capsule (Reineccius, 1995). The only disadvantage of using the freeze drying technique ok that it consumes high energy, processing duration is very long. The obtained structure will be very porous in nature. When the freeze drying is compared to spray drying it is 30 to 50 times expensive (Fanger, 1974).

(vii) Liposome Entrapment

The microcapsule formation by using the liposome Entrapment uses the following techniques such as ultra-sonication, reverse-phase evaporation, microfluidization. The sucrose solution is supersaturated and then dispersed into the clusters to form the agglomerates of the sizes ranging from 10 μm to 10000 μm(Vidhyalakshmi *et al*., 2009). These liposome Entrapment is mainly utilized in the pharmaceutical industries as a drug carrier. The outer layers of the liposome are made up of phospholipids. This phospholipid layer forms a capsule as a strong boundary wall where there are no water interactions. This liposome technique was used in the cosmetics industries further it found applications in pharmaceutical industries. Recently it is being used in food industries majorly in cheese making (New, 1990; Ghychy & Gareiss, 1993; Kirby, 1993). Lecithin obtained from the soybean oil during the purification of oil and phosphatidylcholines isolated from the egg yolk are commonly used forms of phospholipids in the food industries (Martin, 1990). The bilayer formed by the phospholipids cannot be permeated by the sugar or larger polar molecules. This will protect the active ingredient present in the core of liposome known as encapsulated component. The liposomes are manufactured by three different methods viz. lipid formulation, re-dispersion, oxidation and solubilization. The only disadvantage of using the liposome Entrapment is that at the capsule encapsulation healed is very low while it has a limited application due to its chemical and physical instability nature(Shahidi & Han, 1993). Various method of liposome entrapment is used in food industries as per the need of the final product properties. The factors kept in mind during the encapsulation are number of bilayer formed per vesicle, size distribution of vesicle and encapsulation efficiency to be obtained for active ingredient in liposome entrapment.

(viii) Co-acervation

Earlier this method of Co-acervation was used for the production carbon less paper. But now a days this technique finds his wide application in the food industries. Recently the food grade materials are used as the carrier medium. In this method of encapsulation, protein is used as gelling agent. Further the emulsification of the protein and desired essential oil is carried. The polymer solution is used as the coating material which is removed in the liquid form in order to solidify the product. The solidified molecules can be collected by two methods viz. filtration or centrifugation. The Co- acervation is a simple technique of encapsulation which can be carried out by using the colloidal solutions of gelation or gum acacia (Luzzi & Gerraughty, 1964). Such different types of viscous solutions common coating materials used for the encapsulation of essential and flavored oils. By the usage of low temperatures, these coating materials of wall become hard (Thevenet F, 1995). These wall materials are generally hydrophobic in nature while inside core material can be hydrophilic in nature. This makes release of core material delayed. But the release of core material can be obtained by using hot water, high pressure or chemical reaction (Bakan, 1969). The advantage of the method is that it's very efficient method to encapsulate on contrary the only disadvantage of method is being expensive on commercial scale also.

(ix) Inclusion Complexation

The vitamins A, k and E are encapsulated by using this technique of inclusion complexation. Generally for this purpose beta-cyclodextrin is used. The cyclo dextrin possesses a property of slow release of the desired bio-active components which is encapsulated it in. The cyclodextrins are only allowed in japan and eastern Europe for the purpose of food usage (Dziezak, 1988). The center of the encapsulated molecule is hydrophobic in nature, while hydrophobic material is used at the outer surface layer. The polar molecules are replaced by the less polar molecules (Risch, 1995). Water acts as suspension medium in this method of encapsulation.

(x) Co-Crystallization

The matrix of sucrose is used for the entrapment of the active ingredient is called as co-crystallization. The crystals of sizes 3 to 30μm are produced spontaneously in the encapsulation technique of co-crystallization. By using this technique different types of food ingredients can be entrapped in the sucrose matrix and can be converted to crystalline form. Due to which the stability of the active components of the food can be achieved as well as properties are enhanced with increased shelf life. This converted active ingredients/ component provides very useful and interesting characteristics which are applicable in the production of various products in the food processing industries. In the co-crystallization process, sucrose is the carrier material for encapsulation. Hence the syrup of sucrose is maintained at a high temperature in a super saturated state to prevent crystallization of sucrose. The stipulated/ calculated amount of the core material is added to the sucrose syrup. The transformation of the material begins at a specific temperature. To extent and promote the nucleation process continues agitation is provided. After completing the nucleation and agglomeration process is over, the material is transferred to another vessel for drying process. Further the encapsulated material is sieved by using different screens to obtain the proper segregation of desired and uniform size of the encapsulated material. Since the encapsulation process used three major steps of nucleation, agglomeration and crystallization; Heat is generated during the crystallization process. Which is effectively utilized by the encapsulated particles for the drying and dehydration as heat of crystallization. This is one of the advantages of the co-crystallization process where the dried powder form of encapsulated material is obtained without any addition drying and supply of heat. Which will saved the energy required for the drying process.

8.3 Practical Applications

The integrity of the food ingredients viz. vitamins, salts, water, minerals and bioactive compounds can be achieved by using the microencapsulation technology. This can be achieved by modifying the nature of the original state of the compound. A core material can be captured in the shell or coating where, the science is called as microencapsulation. An increased shelf life can be obtained of the food products when we use microencapsulation. The micro encapsulated product may add an appealing taste and mask the off taste of the nutrients and make them more palatable. A key role can be played by the microencapsulation in maintaining the well ness of the foods, keeping the nutraceutical values and the bioactive compounds in the functional foods without sacrificing the smell, taste and convenience. These developed products can provide a large scale of purchase points in the market.

8.4 Conclusion

Microencapsulation is one of the effective protection technique which provides the protection for the active ingredients in the food product. The oxidation, evaporation and migration of these active compounds from the foods can be restricted by using the micro encapsulation. This technology of micro encapsulation can be utilized for the development of the high quality functional products which will possess improved functional and physical properties. It has a wide range of application in the field of pharmaceuticals, nutraceuticals and cosmetic industries. There are numerous techniques reviewed in this paper which are used for the manufacturing of the encapsulating the active food ingredients viz. extrusion, spray drying, spray chilling, fluidized bed coating, liposome entrapment, lypholization, centrifugal coating, emulsions, crystallization, spray cooling, rotational suspensions etc. Despite of its advantages, it has a limited use in the food industry due to its high cost of production, which will eventually increase the product cost. Now its challenge for the food technologists to make the micro encapsulation technique on the commercially viable scale with affordable prices.

8.5 References

Abbas, S., Da Wei, C., Hayat, K., and Xiaoming, Z. 2012. Ascorbic acid: microencapsulation techniques and trends—a review. Food Reviews International, 28(4), 343-374.

Anjani, K., Kailasapathy, K., and Phillips, M. 2007. Microencapsulation of enzymes for potential application in acceleration of cheese ripening. International Dairy Journal, 17(1), 79-86.

Bakan JA. 1969. National Cash Register. US Patent No. 3,436, 355.

Brown, E. N., Kessler, M. R., Sottos, N. R., and White, S. R. 2003. *In situ* poly (urea-formaldehyde) microencapsulation of dicyclopentadiene. Journal of microencapsulation, 20(6), 719-730.

Blenford D. 1986. Fully protected. Food Flavor Ingred. Proc. Pckg. July, p. 43.

Dziezak JD. 1988. Microencapsulation and encapsulated food ingredients. Food Technol. 42, 136-151.

Desai, K. G. H., and Jin Park, H. 2005. Recent developments in microencapsulation of food ingredients. Drying technology, 23(7), 1361-1394.

F. Gibbs, Selim Kermasha, Inteaz Alli, Catherine N. and Mulligan, B. 1999. Encapsulation in the food industry: a review. International Journal of Food Sciences and Nutrition, 50(3), 213-224.

Fang, Z., and Bhandari, B. 2010. Encapsulation of polyphenols–a review. Trends in Food Science and Technology, 21(10), 510-523.

Fanger, G. O. 1974. Microencapsulation: a brief history and introduction. In Microencapsulation Springer, Boston, MA, 1-20.

Flinn, J. E., and Nack, H. 1967. Advances in microencapsulation techniques. Battele Technical Review,16(2), 2.

Frascareli, E. C., Silva, V. M., Tonon, R. V., and Hubinger, M.D. 2012. Effect of process conditions on the microencapsulation of coffee oil by spray drying. Food and bioproducts processing, 90(3), 413-424.

Garti, N., and McClements, D.J. (Eds.). 2012. Encapsulation technologies and delivery systems for food ingredients and nutraceuticals. Elsevier.

Goud, K and Park, H.J. 2005. Recent Developments in Microencapsulation of Food Ingredients. Drying Technology. 23: 1361-1394.

Gouin, S. 2004. Microencapsulation: industrial appraisal of existing technologies and trends. Trends in food science & technology, 15(7-8), 330-347.

Guzey, D., and McClements, D. J. 2006. Formation, stability and properties of multilayer emulsions for application in the food industry. Advances in Colloid and Interface Science, 128, 227-248.

Ghychy M and Gareiss J. 1993. Industrial preparation ofliposomes for cosmetic products. Presented at a Workshopon the Controlled Delivery of Consumer Product s, Controlled Release Society, 21-23 October, Geneva.

Kondo, A. 1980. Microencapsulation utilizing phase separation from an aqueous solution system, in Microcapsule Processing and Technology, van Valkenburg, J. W., Ed., Marcel Dekker, New York, 70.

Kirby CJ. 1993. Controlled delivery of functional food ingredients: opportunities for liposomes in the food industry. In Liposome Technology, 2nd ed. Vol. 2: Entrapment of Drugs and Other Materials, ed. G Gregoriadis, Boca Raton, FL: CRC Press pp. 215-232.

Luzzi LA and Gerraughty RJ. 1964. Effects of selected variables on the extractability of oils from coacervate capsules. J. Pharm. Sci. 53, 429-435.

Martín Banderas, L., Flores Mosquera, M., RiescoChueca, P., Rodríguez Gil, A., Cebolla, Á., Chávez, S., and GañánCalvo, A. M. 2005. Flow focusing: a versatile technology to produce size controlled and specific-morphology microparticles. Small, 1(7), 688-692, .

Martin FP. 1990. Pharmaceutical manufacturing of lipsomes. In Specialized Drug Delivery Systems, ed. P Tyle, New York: Marcel Dekker pp. 50-57.

New RRC. 1993. Preparation of lipsomes. In Lipsomes, a Practical Approach, ed. RCC New, NewYork: IRL Press pp. 33-104.

ONO, F. 1980. New encapsulation technique with protein-carbohydrate matrix. Nippon Shokuhin Kogyo Gakkaishi, 27(10), 529-535.

ONO, F., and Aoyama, Y. 1979. Encapsulation and stabilization of oily substances by protein and carbohydrate. Nippon Shokuhin Kogyo Gakkaishi, 26(1), 13-17.

Oxley, J. 2014. Overview of microencapsulation process technologies. In Microencapsulation in the food industry, 35-46.

Reineccius, G. A., and Bangs, W. E. 1985. Spray drying of food flavors. III. Optimum infeed concentrations for the retention of artificial flavors. Perfumer and Flavorist, 9, 27-29.

Reineccius, G.A. Liposomes for controlled release in the food industry. In Encapsulationand Controlled Release of Food Ingredients, American Chemical Society Symposium Series no. Washington, DC: American Chemical Society, 113–131, 1995.

Risch SJ. 1995. Encapsulation: overview of uses and techniques. In Encapsulation and Controlled Release of Food Ingredient, eds SJ Risch & GA Reineccius, ACS Symposium Series 590. Washington, DC: American Chemical Society. pp. 1-7.

Shahidi, F., and Han, X. Q. 1993. Encapsulation of food ingredients. Critical Reviews in Food Science and Nutrition, 33(6), 501-547.

Sparks, R.E. 1989. Microencapsulation. Encyclopedia of Chemical Processing and Design. Marcel Dekker, New York, 163-180.

Thevenet F. 1995. Acacia gums: natural encapsulation agent for food ingredients. In Encapsulation and Controlled Release of Food Ingredient, eds SJ Risch& GA Reineccius, ACS Symposium Series 590. Washington, DC: American Chemical Society. pp. 51-59.

Taylor AH. 1983. Encapsulation systems and their applications in the flavor industry. Food Flavol. Ingred. Proc.Pckg., Sept., pp. 48-52.

Vidhyalakshmi, R., Bhakyaraj, R., and Subhasree, R.S. 2009. Encapsulation: the future of probiotics-a review. AdvBiol Res, 3(3-4), 96-103, 2009.

9

Malting and Its Effect on Nutritive Value of Food

***Rakhi Singh**, *Vandana Kaushal*, *S. Thangalakshami*
and *Anurag Singh***

ABSTRACT

Malting is the process accompanied by restricted sprouting of cereals in humid atmosphere along with controlled set of conditions. It is a process of soaking, germination and drying that causes changes to the microstructure of grains. Malting causes changes in functional and nutritive value of grains. Malting results in increase in protein content, mineral availability, in vitro protein digestibility, oil absorption capacity, water absorption capacity, water solubility index and fall in level of antinutritional factors, viscosity and bulk density. The malting conditions are significantly affected by varietal difference, maturity of grains and structural composition. Therefore, optimum malting conditions should be determined for grains in order to avail the health and nutrition benefits.

Keywords: Malting, cleaning, steeping, germination, drying, klining

9.1 Introduction

There are many traditional methods for preparing food and malting is one of them. Malting increases the overall nutritional profile and reduces the antinutritional factors. Malting is termed as controlled germination followed by controlled drying as it results in desirable physical and chemical changes in grains which are further stabilized by drying. Development of hydrolytic enzymes is the main objective of malting which results in modification of structural integrity and increased availability of nutrients. Carciochi *et al.* (2016a) studied the effect of roasting on green quinoa malt and concluded malting accompanying a moderate thermal treatment can be contemplated as an effective process to enrich antioxidant compounds, phenolic level and maillard reaction products which can be further used as a functional ingredient.

*Corresponding Author

While working on quinoa flour (Aguilar *et al.*, 2019)concluded malting process results in increase antioxidant capacity, phenolic compounds and flavonoids, furthermore reducing sugars andascorbic acid, hence it could be used in formulating novel nutritive products. Motta *et al.*, (2019) while working on steaming, boiling and malting concluded malting resulted in higher nutritional profile such as protein content and essential amino acids values as compared to raw, boiled and steamed buckwheat, amaranth and quinoa. Malting is well known to improve the nutritional and nutraceutical value of cereals (Subba Rao & Muralikrishna, 2002b) and it's a great alternative to improve absorption of bound phenolic in human.

Malting process consists of four basic unit operations namely

1. Precleaning
2. Steeping
3. Germination
4. Kilning (Drying)

9.2 Precleaning

Precleaning step includes cleaning, grading and storage.Firstly sieving is done followed by removal of straw, leaves, earth and other light-weighted materials by passing grains through air (Owens, 2001). Cleaning is done to remove foreign materials based on difference in their physical properties. After harvesting grains are prepared for storage under dormant condition prior to malting to prevent insect infestation. Moisture content plays an important role in insect infestation, therefore, it is recommended to keep moisture between 10% and 12% for prolonged storage and initial moisture down to 15% to 16% for short duration storage. The fine particles are removed with the help of cleaners and separators and destoner may be equipped if necessary (Guido & Moreira, 2013).

9.3 Steeping

Grains are soaked in water to imbibe enough water to swell. Initially respiration increases slowly then rapidly to cause accumulation of CO2 and heat further enhanced by microflora of steeped grains. The steeping stage comprises of alternate intervals of grain immersed in water and interval with water drained from the grain. Former is known as —under water periods and latter is known as –air rest periods, whereas combination of both maintains and enhances the germination efficiency (Guine *et al.*, 2016). Hydration of grains makes embryo active inside kernel and hydrated embryo utilizes oxygen for respiration. The dissolved oxygen is readily absorbed by organic material and microflora in the grain during the underwater period and in case of unavailability of water for long duration, respiration with release of carbon dioxide may induce alcoholic fermentation. Steeping also assists in removal of dust layer and adhering microorganisms. Husk components leaches in steep water that may hinder the germination and possibly may hamper the proper flocculation of yeast (Guine *et al.*, 2016).

While steeping huge amount of water is utilized. The water consumption during steeping can be reduced particularly by spray-steeping instead of sequential-steeping as well as recirculating chill and humidifying water but reuse of steep out water results in slow germination due to occupancy of germination inhibitors. Thus membrane bioreactors are installed for eliminating germination inhibitors (Guiga *et al.*, 2008).

9.4 Germination

Germination is distinguished by growth of embryo, demonstrated by growth of rootlets and increase in length of sprout(acrospires) along with auxiliary modification in the endosperm content. When grains soak enough water during steeping, the metabolic process of germination starts causing a large increase in enzyme production and starch hydrolysis, these hydrolytic enzymes are found in scutellum and outermost layer of grain.

The modification of the starch in endosperm is initiated owing to the enzymatic degradation and accordingly reduction in their strength can be observed. Resultantly dry malt is more fragile and easily crushed compared to dry grain. Gibberellin the growth promoting hormone is released by enzymes and the scutellum and it results in further releases of various hydrolytic enzymes from aleurone layer. Gibberellin is responsible for sole production and release of α-Amylase and majority of α-Amylase enzyme originates from the aleurone layer and some minor amount from the scutellum layer. Non-uniform malt modification is caused by irregular germination an dgibberellin synthesis resulting in irregular synthesis and release of hydrolytic enzymes (Palmer 1994).

Generally, ―Chitted grains are preferred for malting that have the crown of the rootlet emerging through the husk. Optimum malt is obtained when required changes for a specific type of malt is obtained and malting loss is minimum However, malting loss resulting from respiration is inevitable and depending on malt cultivar ranges 4 % to 8 %.

It is proposed that variations in malt modification are also attributed to variation in grains‘ protein content and different types of proteins hampering enzymatic changes in malting (Palmer 2000). The metabolism of the germinating grain is an exoergic process. Rate of germination and its modification intensity depends on regulating temperature and moisture content of the kernel. Moisture content, temperature, germination time, and oxygen availability plays crucial role in germination. Germination time varies from 2-6 days depending on the type of grain and temperature ranges from 20°C and 30°C. Germination leads to the synthesis of various endogenous hydrolytic enzymes in germ and these enzymes brings structural changes and breaks down complex molecules for utilization by growing embryo (Baranwal, 2017).

9.5 Drying

The final stage of malting is drying performed at 50°C for 24 hours and is needed to stop the further growth of kernels, reduce the moisture content and water activity and reduce the viable number of microbes. Therefore, produces a shelf-stable product with high nutritional and flavor characteristics. High hygroscopicity and soluble nutrient content of malt make them prone to microbial spoilage therefore after kilning low moisture condition should be maintained (Bokulich & Bamforth, 2013) . Drying is followed by removal of shoots and root then milling is done of malted kernels to form malted flour which can be further utilized for preparing varieties of food products.

9.6 Microbiology of Malting

The viable number of microorganisms starts increasing during steeping and continues throughout the germination period, however the few viable number of microorganisms are washed away with steep water and few microbial reductions occur during kilning but overall there is jump in microbial count compared to original grain(Douglas & Flannigan, 1988; Noots *et al.*, 1999; Petters *et al.*, 1988).Growth of toxic microorganism are deleterious to malt quality and few microorganisms even compete for oxygen in steeped liquor further hampering germination(Briggs & McGuinness, 1993; Laitila *et al.*, 2006), rootlet growth and breakdown of starch. Additionally few plant hormones (indole-3-acetate in vitro, gibberellic acid and abscisic acids) are certainly produced by few bacteria and fungi isolated from barley affecting germination and enzyme production. Few steps such as changing the steep liquor between resting period of steeping, reducing dissolved nutrients level and constant optimum steep temperature are the measure to reduce the repressing effect of microbial growth. Few researchers have suggested the inoculation of microbes to steep water since they will compete with detrimental microflora to control their growth during germination. Such as LAB starter culture used as inoculant for malting to improve safety and quality. It is a biological control method, used to control the growth of spoilage microorganism or toxigenic fungi sp. Fusarium (Bokulich & Bamforth, 2013).

Study done by Mohammed *et al.* (2017) on wheat and millet found that by the end of the steeping phase the viable bacteria count reduced in both wheat and millet . This may be attributed to accumulation of the organic acid, decrease oxygen tension and few other toxic compounds as well as low pH value(Table 1). However, proliferation in the fungal microflora during steeping was observed, resulting in a bacterial coating. Increase in yeasts and mould count during steeping in millet and in wheat is an indication of their potential contamination during storage as studied by(Petters *et al.*, 1988) on maize. Due to the ability of bacteria to form matrix composite in form of biofilm during germination (Douglas & Flannigan, 1988), bacteria was the dominant microflora in the sample. Unlike during steeping, yeasts and mould count picked up during germination. An increase in fungal count observed during the germination of both samples may be attributed to low water activity, creating a favorable environment for their proliferation (O'Sullivan *et al.*, 1999). Kilning has a huge impact on microbial stability of the malt, since it results in thermal degradation by kilning in temperature range of 75 $^{\circ}$C -85°C for 24h. Few heat-resistant bacteria may be present due to their ability to form spores under critical conditions.

Lactic acid bacteria were the primary microorganism during steeping. Due to the low pH tolerance, Leuconostoc species were found and others such as Bacillus species which is important producers of extracellular proteases. While presence of others genera such as staphylococcus sp was an indication of possible contamination from environment which is in consistent with study of (Noots *et al.*, 1999) . Nevertheless, great variation in species diversity was observed between different malting periods. Unlike during steeping, yeasts have found several genera in cereals including Saccharomyces sp and Candida are dominant at the end of germination.

Table 1: Microbiological count in malting (Mohammed *et al.*, 2017) (* mean value)

Cereals	Aerobic bacterial count	Aerobic mould count
Wheat	1.08 x 104cfu/ml *	1.00 x 104cfu/ml*
Millet	1.88 x 103cfu/ml *	5.6 x 102cfu/ml*

The results in table 2 revealed bacteria and mould isolated from wheat and millet samples at different malting periods where Bacillus sp and Staphylococcus sp were the predominant bacteria isolated within 0- 144 hours and 24-96 hours of malting period respectively and predominant fungi species isolated at 0-144 hours of malting were Aspergillus and Penicillium sp at 24-48 hours of the malting period.

Table 2: Microorganisms isolated from wheat and millet samples at different malting periods (Mohammed *et al.*, 2017)

Malting period (hr)	Probable microorganisms in millet	Probable microorganisms in Wheat
0	*Salmonella* sp *Penicillium* sp	*Escherichia coli* *Klebsiella* sp *Aspergillus* sp.
24	*Escherichia coli* *Staphylococcus* sp *Aspergillus* sp.	*Streptococcus* sp *Staphylococcus* sp *Penicillium* sp.
48	*Streptococcus* sp *Leuconostoc* sp *Aspergillus* sp.	*Streptococcus* sp *Leuconostoc* sp *Aspergillus* sp.
72	*Staphylococcus* sp *Bacillus* sp *Mucor* sp	*Bacillus* sp *Leuconostoc* sp *Rhizopus* sp
96	*Lactobacillus* sp *Staphylococcus* sp *Penicillium* sp. *Saccharomyces* sp	*Bacillus* sp *Lactobacillus* sp *Penicillium* sp.
144	*Bacillus* sp *Clostridium* sp *Candida* sp	*Bacillus* sp *Clostridium* sp *Saccharomyces* sp

9.7 Effect of Mating on Functional Properties and Nutritive Value of Food

Malting affects various functional and nutritional properties of the grains as given in fig.1 and various studies conducted in this regard are summarized in table 3.)

(i) Functional Property

Functional properties, such as solubility, gelation, viscosity, water and fat binding properties, reflect the level of protein interaction with water, while relations between protein and fat is determined by fat absorption and emulsion. Ojha *et al.* (2018) reported the improved functional property of sorghum flour with malting and fermentation

(ii) Bulk Density

Break down of starch during malting reduces starch content and decreases the bulk. (Ojha *et al.*, 2018) reported 8.64 % decrease in bulk density by malting of sorghum flour. Deepali *et al.*, (2013) studied the functional properties of little, barnyard and foxtail millet and reported the highest reduction in bulk density of the Malted kutki (little millet) by 30% followed by 29% in case of sanwan (Barnyard millet) it was and lowest in kangni (foxtail millet) 20%. After germination, devegetation and drying of malted grains of millet attributed to reduction in the bulk density.

(iii) Viscosity

Due to production of higher soluble components by the action of amylase and microbes during germination may attribute to decrease viscosity of malted product (Uvere *et al.*, 2002). Ojha *et al.*, (2018) reported the decrease in viscosity of whole sorghum flour when subjected to malting and fermentation. Decrease in viscosity reported to be 10.63 % and 12.76 % during malting and fermentation compared to whole sorghum flour, respectively. Grossmann *et al.* (1998) suggested the three day germination at 20-25°C for corn kernels to obtain a low viscosity and higher level of proteins for nutritional interest. Obatolu & Cole (2000) reported the same trend of decrease in viscosity of the composite blends by adding soybean. The reduction in flow of malted products is attributed to starch degradation into simple molecules by the action enzyme amylase.

Presence of high amount of starch contributes to high pasting viscosity. Yang *et al.* (2020) Reported decrease in pasting property of wheat flour by incorporating malted wheat due to the breakdown of starch by enzymatic action while malting.

(iv) Absorption Capacity

Physical capturing and binding of fat to protein molecules results in fat absorption (Wang & Kinsella, 1976). The increase in oil absorption capacity is vital property as it may assist in maintaining and improving the mouth feel of food products for various application such as meat extenders (C *et al.*, 2013). Dispersability and segregation of protein increases amount of lipophilic constituent while germination thus, increases oil absorption capacity (Deepali *et al.*, 2013). Ojha *et al.* (2018) reported 82% increase in oil absorption capacity of malted flour whereas in whole sorghum flour was found to be 68%.

(v) Water Absorption

Nature as well as reaction between protein and carbohydrate plays crucial role in water absorption capacity (Echendu *et al.*, 2005; McWatters *et al.*, 2003). Obatolu & Cole (2000) also reported the same trend of increased water absorption capacity in flour blends of cowpea and soybean in malted maize by 2.94 to 5.40 %.Lorenz & Kulp (1981) observed an increase of starch swelling capacity in long germination duration, attributed to action of amylases enzyme in crystalline regions of the granule, increasing hygroscopicity. Factor probably contributing to the increase of water absorption in dexin flour is majorly the particle area of flour, which decreases with germination time, as a consequence of breakdown of cell wall of kernel. Water solubility index is an important characteristic for cereal flours used in beverages preparation. An increase in water solubility was reported by(Grossmann *et al.*, 1998) by increasing the temperature and time of germination. The maximum value was 13.29%, which represented a 400% increase in water solubility index compared to the control. Genesis of lower molecular weight compounds due to the activity of amylases and proteases could be the result for observed trend (Lorenz,1980; Lukov & Bushuk,1984).Ability of starch granules to swell freely during heating in glut of water is referred as swelling power. Singh *et al.* (2017) observed crucial role of various germination parameters on the swelling power of sorghum flour. Un-malted sorghum flour resulted in the highest swelling power (9.77 g/g) and it decreased remarkably with the germination time and temperature. Germination causes accelerated release of starch and fiber degrading enzymes resulting in high level of dextrin and fermentable sugars by hydrolyzingstarch. Released sugars further evoke and establish crosslink in the amorphous region between the starch chains ,thus making starch unavailable for binding with water (Cornejo & Rosell, 2015; Peroni-Okita *et al.*, 2013). Nelson-Quartey *et al.* (2007) also reported the same trend of reduced swelling power from 11.7% to 7.7% by malting of maize. Yang *et al.* (2020) reported 1.0% substitution of malted wheat flour to untreated wheat flourincreased the water adsorption to 63.6%, however further replacement with malted wheat flour decreased the water absorption to 60.1%. The observed trend could to attribute to decrease in available gluten to form gluten network by its conversion into peptides and amino acids. Unavailability of gluten also results in decrease in extensibility of prepared dough for cookies.

9.8 Nutritive Value

The effect of germination on energy and nutrient density of sorghum-based complementary foods was studied by Tizazu *et al.* (2010) and it was found that germination increased significantly contents of crude protein from 12.25% and 10.44% to 12.65% and 10.87% for two varieties of sorghum respectively. In another study performed by Najdi Hejazi *et al.* (2016) on amaranth grains was found germination resulted in 8% increase in the availability of digestible protein content after 1.5 day of germination at 26 °C.IVDP was increased to around 84%. Study on acha grains by Lasekan *et al.* (2010) found that protein content was affected by germination time and drying time.

Vast studies have shown the ability of germination to improve the protein content. During germination, owing to the absorption of nitrates and further metabolism of nitrogenous compounds from carbohydrate store by nitrates may result in increased crude protein content (Morgan *et al*., 1993). Additionally, release of storage proteins mainly prolamine into albumins and globulins and accordingly transformation of glutamic and proline amino acids into the limiting amino acids such as lysine caused by germination(Chavan & Kadam, 1989). Study by Najdi Hejazi *et al.* (2016) reported the germinated seeds yield higher protein content (14.86 g/100 g) and digestion (83.58%) as compared to the native grains.

Mohammed *et al.* (2017) found that there was slight decrease in protein content in sample of un- malted and unfermented blend of cereals as compared to sample having malted and fermented blend. This could be as a result of accumulation of stored nitrogen in wheat and millet to aid germination. Adebiyi *et al.* (2017a) reported the fiber, carbohydrate and energy values of fermented and malted samples were much highwith low amount of ash content. The observed reduction in ash content might be due to the extraction of soluble inorganic salts while fermentation and malting processes.

Change in acidity value was observed by Mohammed *et al.* (2017)during malting period from 0 to 144 hours in millet and significant decrease in acidity at various malting periods were observed owing to the hydrolysis of organic molecules to fatty acids, lactic acid, acetic acid, amino acids and phosphate as supported by study of Uvere *et al.* (2010) .

The effects of fermentation and malting on pearl millet by Adebiyi *et al.* (2017) showed essential amino acid content significantly increased during malting and fermentation. Arginine, Histidine and HO-Proline were 2–3 times higher in malted pearl millet flour and have been ascribed to improved activities of hydrolytic enzymes and degradation of complex molecules.

Study by Lekjing and Venkatachalam (2020)also supports the above discussion while working on HomChaiya rice. Higher kilning temperature also contributes to gradual increase in the levels of few amino acids. Nie *et al.* (2010) observed changes in the amino acid composition of barley malt from extended germination as well as from kilning temperature.

With respect to fat content, Adebiyi *et al*., (2017b) comparing the nutritional quality and sensory acceptability of fermented and malted flour biscuits of pearl millet reported the crude fat content of the flours between 1.7%– 2.3%, and it was lower than the unmalted flour. The comparatively lower values of fat obtained from malted and fermented flour samples could be attributed to the enzymatic degradation of lipids during fermentation and malting. The added fat contributed to the high fat in biscuit samples (Caponio *et al*., 2008) .Similar trend of decrease in fat content by malting was reported by Mohammed *et al.* (2017) who worked on millets enriched with locust beans. During germination process, starch is broken down into low molecular weight carbohydrates (oligoand disaccharide) by the production of infant foods, weaning foods and enteral foods. Malted millet flour can also be activity of amylase enzymes. Resulted germinated flour had reduced water holding capacity and high energy density which enhance its potential in the used in the production of various other items such as milk-based beverages, confectionaryand cakes (Shobana *et al.*, 2013).

9.9 Anitnutritonal Factors

Antinutritional factors are generally toxic and impair protein digestibility and mineral availability. Study by Najdi Hejazi *et al.* (2016) demonstrated reduction in phytic acid and oxalate content after germination while an increase in the tannin. Germination for 48 h at 30°C decreases phytic acid to the greatest level of 30%.

9.10 Mineral Availability

A major hurdle to the availability of minerals is the presence of harmful and antinutrient substances. Study on malting and fermentation of pearl millet flour by Adebiyi *et al.* (2017b) illustrated that malting resulted in increased mineral availability such as Ca, Cu, Fe, Mg, Mn, P, Na and Zn. Malting has been shown to improve the extractability of minerals via enzyme synthesis and cell wall solubilization (Krishnan *et al.*, 2012; Sade, 2009). Biscuits from malted and fermented flours had comparatively less mineral contents except for Ca and Fe, indicating the heat stability of both minerals. Overall the mineral content of biscuits prepared from malting has higher mineral content and maillard reaction during baking also contributed to high mineral content owing to formation of complexes or disassociation of compounds affecting the mineral solubility and availability(Delgado- Andrade *et al.*, 2011).

Platel *et al.*, (2010) investigated the effect of malting on the bioaccessibility of iron, zinc, calcium, copper, and manganese in finger millet, wheat, and barley. Malting results in increase bioavailability of iron by more than 3-fold in two varieties of finger millet and by more than 2-fold increase in wheat, whereas no such beneficial result was observed in barley. The bioaccessibility of zinc increased to 234% and 100% in wheat and barley, however it in finger millet. Different trend of increase in mineral availability was noted for finger millet and barley. However, author concluded malting could be employed as appropriate strategy to obtain minerals from food matrix.

9.11 Bioactive Compounds

Various studies have proved that phenolic compounds possess health promoting properties. Results from the study of Adebiyi *et al.* (2017) demonstrated that malted pearl millet flour had the highest total phenolic content ,74.05mg GAE/g and total flavonoid content,39.80mg CE/g. Perpetual production and polymerization of phenolic compounds occurs during the germination (Jood *et al.*, 1987; Taylor & Duodu, 2015)and loss of dry matter and potential release of condensed procyanidins occurring throughout the malting process probably accord to the increase in total phenolic content(Taylor & Duodu, 2015) .

Udeh et al., (2018a) reported increase in the concentration of catechin, epicatechin, and protocatechuic acid following 3rd and 4th day of malting finger millet and sorghum malt and complete loss of taxifolin and hesperitin. Subba Rao & Muralikrishna (2002a) reported protocatechuic, gallic, and caffeic acids are the major free phenolic acids and a three times fall in protocatechuic acid content, whereas infinitesimal was caffeic acid following 96 h of malting. However, other free phenolic acids such as gallic, vanillic, coumaric, and ferulic acids increasedand a two times decrease was observed in major bound phenolic acidsupon 96 h of malting.

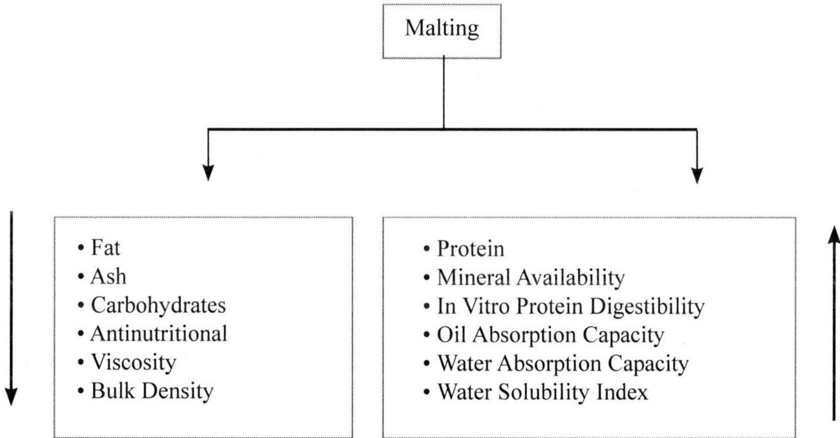

Fig.1: Changes in functional and nutritive value during malting

Table 3: Effect of malting on nutritive and functional parameters of product

Product	Effect of Malting	References
Red Sorghum, White Sorghum and Pearl Millet	Malted cereals digestibility ranged between 34.5-68.1% and 19-33% phytate reduction	(Onyango et al., 2013)
Sorghum	Reduction in phytate, tannin, and oxalate by 40%, 16.12% and 49.1%, respectively	(Ojha et al., 2018)
Finger Millet, wheat, and Barley	Increase in iron, zinc ,calcium and manganese bioaccessibility with few exception	(Platel et al., 2010)
Indigenous varieties of foxtail millet	Lower crude fiber content (9.4% and 12.8%, respectively)and Reduction of fat content by 53.3% and 59.7% in red and purple variety	(Choudhury et al., 2011)
Ready to Use complementary food mixes	10.5% protein, fat 24.1% of total energy content and 51.2% of extractable zinc	(Lohia & Udipi, 2015)
Malting of finger millet and amaranth	17% increase in protein availability, 10% increase in total energy and 60% reduction in resistant starch in finger millet For amaranth, germinating resulted in 8% increase in protein availability,11% increase in total energy, 70% reduction in resistant and a 10% increase in the linoleic acid.	(Najdi Hejazi et al., 2016)
Mixed flour of foxtail millet, wheat and chickpea (40:30:30)	Crude fiber, vitamin C, and iron were increased in mixed flour by 13.04%,70%, 26.82%	(Laxmi G et al., 2015)
Quinoa , amaranth and buckwheat: Influence of cooking and malting	Increased total folate content in amaranth by 21% and buck-wheat by 27% and no change in quinoa was observed whereas boiling and steaming reduced the total folate	(Motta et al., 2017)
Malted sorghum flour	Decrease in pasting temperature, peak viscosity, final viscosity, and increase the content of enzyme-susceptible starch and diastatic power	(Hugo et al., 2000)
Wheat malt	15.36 %,12.71 %,41.7% and 59.39 % increase in protein, moisture, iron and phosphorus8.46% and 21.8 % decrease in fat and crude fiber	(St et al., 2019)

Malted wheat flour cookies	10 % substitution of malted wheat decreased water absorption from 62.5 % to 60.1 % Reduced pasting time from 6.12 min to 3.22 min.	(Yang et al., 2020)

9.12 Malting Process Variables

Few malting conditions are accountable for the quality of finished malted products. The control of malting period, temperature of steeping & germination and time of germination &steeping determine nutritive value and various other characteristics of the product.

9.13 Steeping Time

Wijngaard *et al.* (2005)studied the effect of three steeping times were 7 h, 13 h and 80 h on buckwheat quality. Broadly increased moisture content increases malting losses, kolbach index and total beta- amylase activity and reduces total nitrogen, viscosity and brittleness. The desired steeping time can be achieved by maintaining the balance between the desired malt qualities whereas minimizing the malting loss. Malting loss of 7.43%, 7.89% and 10.74% was reported at steeping time of 7 h, 13 h and 80 h. Maximum malting loss of 10.74% exceeds the tolerable range of 6.5% and 10.5% for malted barley and was attributed to higher steeping loss and longer rootlets development. It further results in a drop of total nitrogen content since degraded proteinaceous compounds during germination are stored into rootlets,which are further eliminated during malt cleaning. There is a direct relation between steeping time and steep-in moisture content and total nitrogen was decreased with an increased steeping time. However, progressive degradation in the protein matrix of buckwheat was anticipated to result in increased friability, but unlike the expected pattern friability decreased with an increasing moisture content. Activity of Alpha-amylases and specifically beta-amylases were low in all buckwheat malts leading to starch positive mashes. It was observed that β-amylase exists in a indispersible and dormant form in buckwheat and during malting beta-amylase was dispersible. Overall it was concluded that the recommended moisture content for producing malted buckwheat ranges between 35 and 40%. Montanuci *et al.* (2017) reported levels of α-glucans decreased with increasing hydration period and temperature. Hydration conditions of 10° for 24 h and 20° for 12 h resulted in the highest overall sugar content. Therefore, these parameters were the best treatments to produce malt.

9.14 Malting Period

Udeh *et al.* (2018b) reported 24 h malting attributed to decline in mineral content of the grains except sodium. Malting past 48 h particularly at 96 h resulted in increased mineral content and not resulted in alternation of selenium and cobalt content. The increase in the mineral content after 24h malting could be owing to dissociation and /or reconfiguration of antinutritional components during malting, and also due to the occupancy of mineral enhancers like Vitamin C(Mamiro *et al.*, 2001; Sripriya *et al.*, 1997).

Udeh *et al.* (2018a) reported 72h and 96h of malting time attributed toincrease inquercetin, catechin, epicatechin and protocatechuic acid and exhibited antioxidant activity found to increase with malting period. Varietal difference also playsa crucial role on phenolic content of the malted grains. The reported changes in the antioxidant activity could be attributed to the total phenolic content, released minerals, and other organic compounds with an antioxidant property.

Study conducted by Najdi Hejazi *et al.* (2016)found that germination period resulted in increased protein availability. Initially in-vitro protein digestibility was 76%, and it increased up to around 84% and a maximum value was reported at 28°C for 48h.Phytic acid decreased to the highest level of 30% for germination at 30°C for 48 h while on the other hand, tannin increased throughout germination by maximum 47% compared. It was concluded from the study that while germination duration has notable influence on all parameters namely protein content, in vitro protein digestibility and antinutritional especially oxalate content, germination temperature was solely significant for IVPD, phytic acid and oxalate.

9.15 Germination Time

Study on Hom Chaiya rice by Lekjing & Venkatachalam (2020)exhibited the effect of germination time on malting capability such as germination potential, germination rate and dormancy rate of Hom Chaiya rice. Prolonged germination significantly increased germination rate of malted Hom Chaiya rice. As germination continued, the germination capacity reached supreme level and conversely, the dormancy rate decreased moderately with germination time. The inhibitory influence of husk to cushion the grains from moisture represents dormancy rate and husk is loosened during germination by steeping that imbibes the embryo and induces enzymatic changes in the grain (Carciochi *et al.*, 2016a). Increasing germination duration enhances the diastatic potential of Hom Chaiya rice and gradually increased amylase activities. α-Amylase activities were higher than those of β-amylase in the

Hom Chaiya germinated rice and germination for 7 consecutive days resulted in highest β and α- amylase activities and α-amylase was slightly prevailing a top β-amylase in germinated rice.

Grossmann *et al.* (1998) found decrease in peak viscosity and cold viscosity with increasing germination time till third day and it is desirable for preparing weaning foods having less consistency.

Carciochi *et al.* (2014) outlinedthe increase of up to 101% in phenolic compounds as function of the germination duration in quinoa variety upto 3 day germination. Similar observation were reported for *Chenopodium by* Abderrahim *et al.* (2012).

Singh *et al.* (2017) reported dwindling in swelling capacity with increase in temperature of germination. Value 8.37 g/g after 12 h germination which was slasheddown to 7.75g/g after 24 h and further went down to 6.26 g/g after 48 h of germination. Odo *et al.*, (2016) revealed increased in germination energy with increase in germination time up to day 6 and it indicates the malting potential. By the end of germination, germination energy ranged bewteen,98 % to 99 % and degree of grain maturity contributes to such variation (Hammond & Ayernor, 2001).

9.16 Germination Temperature

Carciochi et al. (2016b)reported the vigour index value was significantly influenced by the temperature of germination. Bois *et al.* (2006) studied the rate of germination on quinoa and found the speed of germination was markedly increased till optimum germination time for cultivar and further increase in temperature could cause heat stress, leading to loss of vigour and delayed germination (Wahid *et al.*, 2007).Najdi Hejazi *et al.* (2016) reported comparatively higher values of protein for the samples germinated at 26°C for germination time of 24, 36, and 48 has compared to 22°Cand 30°C.

9.17 Kilning Temperature

Study by Lekjing & Venkatachalam (2020) demonstrated that increasing kilning temperature resulted in high increase the diastatic potential. Enzyme activity and malting loss of rice gradually surged with kilning temperature whereas, malting yield and viscosity of the samples plunged. Higher kilning temperature increases the protease activity and accordingly soluble protein and unbound amino acids also increase.

9.18 Conclusion

Nutritional profile and functional properties of malted food products are superior compared to their unmalted counterparts due to the lack of endogenous enzymes resulting in high availability of nutrients while decreasing the toxic and harmful substances. Malting results in increase in protein content, mineral availability, in vitro protein digestibility, oil absorption capacity, water absorption capacity, water solubility index and fall in level of fat, ash, carbohydrates, antinutritional factors, viscosity and bulk density. Increase in steeping time results in malting loss and kolbach index. The malting conditions are dependent on each other various other factors such as varietal difference, maturity of grains and structural composition. Therefore, optimum malting conditions should be determined for grains in order to avail the health and nutrition benefit of malting. Further research can be done to evaluate the effect of malting as a pretreatment on novel processing technique for various food products and use of malting technique to prevent diseases by formulating nutraceuticals and other health promoting products.

9.19 References

Abderrahim F, Huanatico E, Repo-Carrasco-Valencia R, Arribas S M, Gonzalez M C, Condezo-Hoyos L. Effect of germination on total phenolic compounds, total antioxidant capacity, Maillard reaction products and oxidative stress markers in canihua (Chenopodium pallidicaule). J.Cereal Sci., 56(2), 410–417, 2012. https://doi.org/10.1016/j.jcs.2012.04.013

Adebiyi J A, Obadina A O, Adebo O A, Kayitesi E. Comparison of nutritional quality and sensory acceptability of biscuits obtained from native, fermented, and malted pearl millet (Pennisetum glaucum) flour. Food Chem., 232, 210–217, 2017. https://doi.org/10.1016/j.foodchem.2017.04.020

Aguilar J, Miano A C, Obregón J, Soriano-Colchado J, Barraza-Jáuregui G. 2019. Malting process as an alternative to obtain high nutritional quality quinoa flour. J.Cereal Sci., 90(June) , https://doi.org/10.1016/j.jcs.2019.102858

Albishi T, John J A, Al-Khalifa A S, Shahidi F. Antioxidative phenolic constituents of skins of onion varieties and their activities. J. Funct. Foods, 5(3), 1191–1203, 2013. https://doi.org/10.1016/j.jff.2013.04.002

Baranwal D. 2017. Malting: An indigenous technology used for improving the nutritional quality of grains- A review. Asian J Dairy Food Res, 36(03),. https://doi.org/10.18805/ajdfr.v36i03.8960

Bois J F, Winkel T, Lhomme J P, Raffaillac J P, Rocheteau A. 2006. Response of some Andean cultivars of quinoa (Chenopodium quinoa Willd.) to temperature: Effects on germination, phenology, growth and freezing. Eur J Agron, 25(4), 299–308, 2006. https://doi.org/10.1016/j.eja.2006.06.007

Bokulich N A, Bamforth C W. The Microbiology of Malting and Brewing. MMBR, 77(2), 157–172, 2013. https://doi.org/10.1128/mmbr.00060-12

Briggs D E, McGuinness G. Microbes on Barley Grains. J. Inst. Brew, 99(3), 249– 255, 1993. https://doi.org/10.1002/j.2050-0416.1993.tb01168.x

Onuegbu N C, Nworah KO, Essien P E, Nwosu J N, Ojukwu M. Proximate, functional and anti-nutritional properties of boiled ukpo seed (Mucunaflagellipes) flour. Int. J. Life Sci, 2(1),1-4,2013.

Caponio F, Summo C, Clodoveo M L, Pasqualone A. Evaluation of the nutritional quality of the lipid fraction of gluten-free biscuits. Euro Food Res Technology , 227(1), 135–139, 2008. https://doi.org/10.1007/s00217-007-0702-0

Carciochi, Ramiro A, Manrique G D, Dimitrov K. Changes in phenolic composition and antioxidant activity during germination of quinoa seeds (Chenopodium quinoa Willd.) . Int Food Research Journal, 21(2), 767,2014.

Carciochi, Ramiro Ariel, Dimitrov K, Galván D´Alessandro L.. Effect of malting conditions on phenolic content, Maillard reaction products formation, and antioxidant activity of quinoa seeds.JFST, 53, 3978–3985,2016. https://doi.org/10.1007/s13197-016-2393-7

Chavan J K, Kadam S S. Nutritional improvement of cereals by sprouting. Crit. Rev. Food Sci. Nutr., 28(5), 401–437,1989. https://doi.org/10.1080/10408398909527508

Cheryan M. Phytic acid interactions in food systems. Crit. Rev. Food Sci. Nutr., 13(4), 297–335, 1980. https://doi.org/10.1080/10408398009527293

Choudhury M, Das P, Baroova B. Nutritional evaluation of popped and malted indigenous millet of Assam. JFST, 48(6), 706–711, 2011. https://doi.org/10.1007/s13197-010-0157-3

Cornejo F, Rosell C M. Influence of germination time of brown rice in relation to flour and gluten free bread quality. JFST, 52(10), 6591–6598, 2015. https://doi.org/10.1007/s13197-015- 1720-8

Deepali A, Anubha U, Sagar Nayak P. Functional characteristics of malted flour of foxtail, barnyard and little millets. AFST, 14(1), 44–49, (2013).

Delgado-Andrade C, Seiquer I, García M M, Galdó G, Navarro M P. Increased Maillard reaction products intake reduces phosphorus digestibility in male adolescent mice. Nutrition, 27(1), 86–91,2011. https://doi.org/10.1016/j.nut.2009.10.009

Dewar J. Influence of malting on sorghum protein quality. Afripro, Workshop on the proteins of sorghum and millet : Enhancing Nutritional and functional properties for Africa,2003. http://www.afripro.org.uk/papers/Paper18Dewar.pdf

Douglas P E,Flannigan B. A microbiological evaluation of barley malt production. J. Inst. Brew, 94(2), 85–88,1988. https://doi.org/10.1002/j.2050-0416. 1988. tb04562.x

Dvořáková M, Guido L F, Dostálek P, Skulilová Z, Moreira M M, Barros A A. Antioxidant Properties of Free, Soluble Ester and Insoluble-Bound Phenolic Compounds in Different Barley Varieties and Corresponding Malts. J. Inst. Brew, 114(1), 27–33, 2008. https://doi.org/10.1002/j.2050-0416.2008.tb00302.x

Echendu C, Onimawo I, Adieze S. Production And Evaluation Of Doughnuts And Biscuits From Maize – Pigeon Pea Flour Blends. Niger. Food J., 22(1), 147–153, 2005. https://doi.org/10.4314/nifoj.v22i1.33580

Goupy P, Hugues M, Boivin P, Amiot M J. Antioxidant composition and activity of barley (Hordeum vulgare) and malt extracts and of isolated phenolic compounds. J. Sci. Food Agric, 79(12), 1625–1634, 1999. https://doi.org/10.1002/(SICI)1097-0010 (199909) 79:12<1625::AID-JSFA411>3.0.CO;2-8

Grossmann M V E, Mandarino J M G, Yabu M C. Chemical composition and functional properties of malted corn flours. Braz Arch Biol Technol, 41(2), 1998. https://doi.org/10.1590/s1516-89131998000200003

Guiga W, Boivin P, Ouarnier N, Fournier F, Fick M. Quantification of the inhibitory effect of steep effluents on barley germination. Process Biochem, 43(3), 311–319, 2008. https://doi.org/10.1016/j.procbio.2007.12.001

Guine R de P. F., Correia, P. M. dos R., & Correia, P. M. dos R. (2016). Engineering Aspects of Cereal and Cereal-Based Products. In Engineering Aspects of Cereal and Cereal-Based Products. CRC Press. https://doi.org/10.1201/b15246

Hammond T K, Ayernor G S. Characteristics of malted rice for the production of sugar syrup. JGSA, 3(3) , 2001. https://doi.org/10.4314/jgsa.v3i3.17771

Hingade ST, Chavan VR, Machewad GM, Deshpande HW.Studies on effect of malting on physiochemical characteristics of wheat malt and barley malt used for preparation of probiotic beverage. J Pharmacogn Phytochem, 8(1), 1811–1813, 2019.

Hough J S, Briggs D E, Stevens R,Young T W(Ed), Malting and Brewing Science, Springer US, 1982. https://doi.org/10.1007/978-1-4615-1799-3

Hugo L F, Rooney L W, Taylor J R N. Malted sorghum as a functional ingredient in composite bread. Cereal Chem, 77(4), 428–432, 2000.

https://doi.org/10.1094/CCHEM.2000.77.4.428

Jood S, Chauhan B M, Kapoor A C. Polyphenols of chickpea and blackgram as affected by domestic processing and cooking methods. J. Sci. Food Agric, 39(2), 145–149, 1987. https://doi.org/10.1002/jsfa.2740390207

Krishnan R, Dharmaraj U, Malleshi N G. Influence of decortication, popping and malting on bioaccessibility of calcium, iron and zinc in finger millet. LWT-Food Sci Technol., 48(2), 169–174, 2012. https://doi.org/10.1016/j.lwt.2012.03.003

Laitila A, Wilhelmson A, Kotaviita E, Olkku J, Home S, Juvonen R. Yeasts in an industrial malting ecosystem. J. Ind. Microbiol Biot., 33(11), 953–966, 2006. https://doi.org/10.1007/s10295-006-0150-z

Lasekan O, Salva J T, Abbas K. Effect of malting conditions and quality characteristics of malt and roasted malt extract from —achal grains (Digitaria exilis Stapf). J. Sci. Food Agric, 90(5), 850–860, 2010. https://doi.org/10.1002/jsfa.3895

Laxmi G, Chaturvedi N, Richa S. The Impact of Malting on Nutritional Composition of Foxtail Millet, Wheat and Chickpea. J Nutr Food Sci, 5(5), 407, 2015. https://doi.org/10.4172/2155- 9600.1000407

Lekjing S, Venkatachalam K. Effects of germination time and kilning temperature on the malting characteristics, biochemical and structural properties of HomChaiya rice. RSC Advances, 10(28), 16254–16265 ,2020. https://doi.org/10.1039/d0ra01165g

Lohia N, Udipi SA. Use of fermentation and malting for development of ready-to-use complementary food mixes. Int J of Food and Nutr Sci, 4(1), 1–4. 2015

Mamiro PRS, Van J, Mwikya SM, Huyghebaert A. In vitro extractability of calcium, iron, and zinc in finger millet and kidney beans during processing. J of Food Sci, 66(9), 1271–1275, 2001. https://doi.org/10.1111/j.1365-2621.2001.tb15200.x.

McWatters KH, Ouedraogo JB, Resurreccion AVA, Hung YC, Phillips RD. Physical and sensory characteristics of sugar cookies containing mixtures of wheat, fonio (Digitaria exilis) and cowpea (Vigna unguiculata) flours. Int J of Food Sci and Tech, 38(4), 403–410, 2003. https://doi.org/10.1046/j.1365-2621.2003.00716.x.

Mella ON. Effects of malting and fermentation on the composition and functionality of sorghum flour. https://digitalcommons.unl.edu/foodscidiss/12, 2011.

Mohammed SSD, Orukotan AA, Musa, J. Effect of fermentation and malting on some cereal weaning foods enriched with African locust beans. JASEM, 21(5), 911, 2017. https://doi.org/10.4314/jasem.v21i5.17.

Montanuci FD, Ribani M, de Matos Jorge L M, Matos Jorge, RM. Effect of steeping time and temperature on malting process. J of Food Proc Eng, 40(4), 2017. https://doi.org/10.1111/jfpe.12519.

Morgan JV, Hunter RR, & O'Haire R. Limiting factors in hydroponic barley grass production. Proceedings of the 8th International Congress on Soilless Culture, Hunter' s Rest, South Africa, October, 1992, 241–261.

Motta C., Castanheira I., Gonzales GB, Delgado I, Torres D, Santos M, Matos AS. Impact of cooking methods and malting on amino acids content in amaranth, buckwheat and quinoa. J. Food Compos. Anal., 76, 58–65, 2019. https://doi.org/10.1016/j.jfca. 2018.10.001.

Motta C, Delgado I, Matos AS, Gonzales, GB, Torres D, Santos M, Chandra-Hioe MV, Arcot, J, Castanheira I.. Folates in quinoa (Chenopodium quinoa), amaranth (Amaranthus sp.) and buckwheat (Fagopyrum esculentum): Influence of cooking and malting. J. Food Compos. Anal., 64(August), 181–187, 2017. https://doi.org/10.1016/j.jfca.2017.09.003.

Najdi HS, Orsat V, Azadi B, Kubow S.. Improvement of the in vitro protein digestibility of amaranth grain through optimization of the malting process. J.Cereal Sci., 68, 59–65, 2016. https://doi.org/10.1016/j.jcs.2015.11.007.

Nelson-Quartey FC, Oduro I N, Ellis WO, Amagloh, FK. Formulation of an infant food based on breadfruit (Artocarpus altilis) and breadnut (Artocarpus camansi). Acta Hortic, 757, 215–224, 2007. https://doi.org/10.17660/ActaHortic.2007.757.29.

Nie C, Wang, C, Zhou, G, Dou, F, Huang, M. Effects of malting conditions on the amino acid compositions of final malt. Afr. J. Biotechnol., 9(53), 9018–9025, 2010. https://doi.org/10.5897/AJB10.370.

Noots I, Delcour J A, Michiels CW. From field Barley to Malt: Detection and specification of microbial activity for quality aspects. Crit. Rev. Microbiol, 25(2), 121–153, 1999. https://doi.org/10.1080/10408419991299257.

O'Sullivan T F, Walsh Y, O'Mahony A, Fitzgerald, GF, Sinderen D. A Comparative Study of Malthouse and Brewhouse Microflora. J. Inst. Brew, 105(1), 55–61, 1999. https://doi.org/10.1002/j.2050-0416.1999.tb00006.x.

Obatolu VA, Cole AH. Functional property of complementary blends of soybean and cowpea with malted or unmalted maize. Food Chem., 70(2), 147–153, 2000. https://doi.org/10.1016/S0308-8146(99)00248-4.

Odo MO, Okorie PA, Ikegwu OJ, Kalu, MA. Malting potentials of hybrid and Local varieties of rice. AJAFS 4(3), 146–151, 2016.

Ojha P, Adhikari R, Karki R, Mishra A, Subedi U, Karki, TB. Malting and fermentation effects on antinutritional components and functional characteristics of sorghum flour. Food Sci. Nutr 6(1), 47–53, 2018. https://doi.org/10.1002/fsn3.525.

Onyango, CA, Ochanda, SO, Mwasaru, MA, Ochieng, JK, Mathooko, FM, Kinyuru JN. Effects of malting and fermentation on anti-nutrient reduction and protein digestibility of red sorghum, white sorghum and pearl millet. J. Food Res., 2(1), 41, 2013. https://doi.org/10.5539/jfr.v2n1p41.

Owens G. Cereals processing technology. CRC Press. 2001

Peroni-Okita, FHG, Cardoso, MB, Agopian, RGD, Louro RP, Nascimento JRO, Purgatto E, Tavares MIB, Lajolo FM, Cordenunsi, BR. The cold storage of green bananas affects the starch degradation during ripening at higher temperature. Carbohydr. Polym, 96(1), 137–147, 2013. https://doi.org/10.1016/j.carbpol.2013.03.050.

Petters HI, Flannigan B, Austin B. Quantitative and qualitative studies of the microflora of barley malt production. J Appl Bacteriol , 65(4), 279–297, 1988. https://doi.org/10.1111/j.1365-2672.1988.tb01895.x

Phiarais BPN, Wijngaard HH, Arendt, EK. The impact of kilning on enzymatic activity of buckwheat malt. J. Inst. Brew, 111(3), 290–298, 2005. https://doi.org/10. 1002/j.2050-0416.2005.tb00685.x.

Platel K, Eipeson SW, Srinivasan K. Bioaccessible mineral content of malted finger millet (*Eleusine coracana*), wheat (*Triticum aestivum*), and barley (*Hordeum vulgare*). J. Agric. Food Chem., 58(13), 8100–8103, 2010. https://doi.org/10.1021/jf100846e.

Ruiz A de, Bressani R. 1990. Effect of germination on the chemical composition and nutritive value of amaranth grain. Cereal Chem., 67(6),519-522.

Samaras TS, Camburn PA, Chandra SX, Gordon, MH, Ames, JM. 2005. Antioxidant properties of kilned and roasted malts. J. Agric. Food Chem., 53(20), 8068–8074, 2005. https://doi.org/10.1021/jf051410f.

Shobana S, Krishnaswamy K, Sudha V, Malleshi NG, Anjana RM, Palaniappan L, Mohan V. —Finger millet (*ragi, eleusine coracana* L.): a review of its nutritional properties, processing, and plausible health benefits, Adv in Food and NutRes, 69, 1-39, 2013

Singh A, Sharma S, Singh B. Effect of germination time and temperature on the functionality and protein solubility of sorghum flour. J.Cereal Sci., 76, 131–139, 2017. https://doi.org/10.1016/j.jcs.2017.06.003

Sripriya G, Antony U, Chandra TS. Changes in carbohydrate, free amino acids, organic acids, phytate and HCl extractability of minerals during germination and fermentation of finger millet (Eleusine coracana). Food Chem., 58(4), 345–350, 1997. https://doi.org/10.1016/S0308- 8146(96)00206-3.

Subba Rao MVSST, Muralikrishna G. Evaluation of the antioxidant properties of free and bound phenolic acids from native and malted finger millet (ragi, E*leusine coracana* Indaf-15). J. Agric. Food Chem., 50(4), 889–892, 2002. https://doi.org/10.1021/jf011210d.

Taylor J R, Duodu, KG. Effects of processing sorghum and millets on their phenolic phytochemicals and the implications of this to the health-enhancing properties of sorghum and millet food and beverage products. J. Sci. Food Agric, 95(2), 225–237, 2015. https://doi.org/10.1002/jsfa.6713

Tizazu S, Urga K, Abuye C, Retta, N. Improvement of energy and nutrient density of sorghum based complementary foods using germination. Afr. J. Food Agric. Nutr. Dev., 10(8) , 2010. https://doi.org/10.4314/ajfand.v10i8.60875.

Udeh HO, Duodu KG, Jideani AIO. Malting period effect on the phenolic composition and antioxidant activity of finger millet (*Eleusine coracana* l. Gaertn) flour. Molecules, 23(9), 2018a. https://doi.org/10.3390/molecules23092091

Udeh HO, Duodu KG, Jideani, AIO. Effect of malting period on physicochemical properties, minerals, and phytic acid of finger millet *(Eleusine coracana)* flour varieties. Food Sci. Nutr 6(7), 1858–1869, 2018b. https://doi.org/10.1002/fsn3.696

Uvere PO, Ngoddy, PO, Nnanyelugo DO. Effect of amylase-rich flour (ARF) treatment on the viscosity of fermented complementary foods. FNB, 23(2), 190–195, 2002. https://doi.org/10.1177/156482650202300208.

Uvere PO, Onyekwere, EU, Ngoddy, PO. Production of maize-bambara groundnut complementary foods fortified pre-fermentation with processed foods rich in calcium, iron, zinc and provitamin A.J. Sci. Food Agric, 90(4), 566–573, 2010. https://doi.org/10.1002/jsfa.3846.

Wahid A, Gelani S, Ashraf M, Foolad M R. 2007. Heat tolerance in plants: An overview. EEB, 61(3),199–223, 2007. https://doi.org/10.1016/j.envexpbot.2007.05.011.

Wang JC, Kinsella JE. Functional properties of novel proteins: alfalfa leaf protein. J. Food Sci, 41(2), 286–292, 1976. https://doi.org/10.1111/j.1365-2621.1976.tb00602.x.

Wijngaard HH, Ulmer HM, Neumann M, Arendt EK. The effect of steeping time on the final malt quality of buckwheat. J. Inst. Brew, 111(3), 275–281, 2005. https://doi.org/10.1002/j.2050-0416.2005.tb00683.x.

Yang B, Guo M, Zhao Z. Incorporation of wheat malt into a cookie recipe and its effect on the physicochemical properties of the corresponding dough and cookies. LWT-FOOD SCI TECHNOL, 117, 108651, 2020. https://doi.org/10.1016/j.lwt.2019.108651.

10

An Insight of Cold Plasma Processing and Its Application in Food Industry

Drishti Kadian **and** ***Chander Mohan****

ABSTRACT

The commencement of the 21st century has endorsed an extended requirement for safe and wholesome food. The fact is that consumer preferences, awareness and their concern towards food safety have been tremendously increased. Therefore, inactivation of pathogenic microorganisms through any treatment becomes essential for food processors to ensure that food is free from disease causing microorganisms and parallelly unaltered with respect to its organoleptic, functional and nourishing characteristic. Food industries predominantly adopt thermal inputs for the preservation of food. Thermal methods, however, lead to unfavourable changes in organoleptic and nutritional attributes, more particularly proteins and vitamins. Hence, there has been continuous development and advances in finding alternatives to conventional processing. Cold plasma technology is one such technique. Plasma can be defined as a quasi-neutral ionised gas and is more commonly termed as 4th state of matter. Advances in plasma engineering have made it possible to generate plasma at atmospheric pressure which ultimately resulted in broadening the spectra of research on cold plasma at various interfaces of biological sciences. This chapter summarises the overview of cold plasma technology processing, numerous findings related to its application in the area of food decontamination, food fortification, seed germination, milk and milk products and modification of starches.

Keywords: cold plasma, non-thermal, plasma chemistry, fortification, modification, germination, plasma generation, novel

10.1 Introduction

Consumer preferences, awareness and their concern towards food safety have been tremendously increased.Food isundoubtedly associated with several diseases when contaminated with microorganisms. Therefore, elimination,

*Corresponding Author

reduction and inactivation of pathogenic microorganisms through any treatment become essential for food processors to ensure that food is free from disease causing microorganisms and with parallellyunaltered organoleptic, functional and nourishing characteristic (Man, 2005). This is usually accomplished by either thermal or non-thermal processing methods. Among thermal and non-thermal processing methods, thermal inputs for the preservation of food are predominantly adopted by food industries. Thermal treatments, usually, however, lead to unfavourable changes in colour, flavour, texture and nutritional attributes, more particularly proteins and vitamins destruction (Misra *et al.*, 2011). Thermal processing methods include pasteurization, canning, sterilization and many more, whereas non-thermal processing includes ionizing radiation, pulse electric field, high-pressure processing etc. As mentioned earlier, thermal processing has a number of drawbacks, particularly from the nutritional and sensory point. Hence, there has been continuous development and advances in finding alternatives to conventional or thermal processing. Cold plasma technology is one such technique.

The foundation stone of cold plasma was established by Sir William Crookes in 1879, who introduced radiant matter as the most predominant state of matter. Subsequently, International Electrical Congress held in Paris where the term plasma was coined by Irwin Langmuir in 1932 (Siddique *et al.*, 2018).

Plasma is categorized as thermal and non-thermal. When there is thermodynamic equilibrium in electron and other gas species, it is referred as thermal whereas the existence of electron and other gas species in non-equilibrium is referred as non-thermal plasma. During its infancy, low temperature conditions were employed to generate cold plasma which narrowed its practical applications. Continuous research on plasma engineering have made it possible to generate plasma at atmospheric pressure which ultimately resulted in broadening the spectra of exploring cold plasma at several interfaces of biological sciences. This chapter summarises the overview of cold plasma technology processing, numerous findings related to its application in the area of food decontamination, food fortification, seed germination, milk and milk products, modification of starches etc.

10.2 What is Cold Plasma?

Plasma is regarded as the 4th state of matter. In the state change hierarchy, it lies after gas: solid→ liquid → gas → plasma, where each step amounts to an increase in energy of the matter(Misra *et al.*, 2011). Plasma is composed of several constituents including photons, ions, electrons, neutrons, free radical species, excited-state atoms and molecules. Plasma constitutes of two types of species: light species comprises photons, and massive species comprises of the remaining constituents. Ekezie *et al.* (2017) stated that it has an equal number of positive and negative ions due to which the overall net charge on plasma is zero.

Plasma generation can be differentiated as a) thermal and b) non-thermal plasma. The distinction made rely on thermodynamic balance between electrons and ions. In thermal plasma, gas is heated to the high temperature and pressure (>10^5 Pa) to achieve a plasma state. Non-thermal plasma (NTP) generates under low-temperature, and reduced vacuum or atmospheric pressure and further is divided into two parts: quasi equilibrium plasma and nonequilibrium plasma which operates at 100-150°C and <60°C, respectively (Misra *et al.*, 2016). The energy required to transfer from electrons is low than the efficiency of NTP, also known as cold plasma, to cool ions and some massive particles (molecules).

10.3 Plasma Chemistry and Generation

The central and important element for plasma processing is ionization trailing by additional aspects like rate constant, rate of reaction, free path available and distribution of electron energy (Fridman, 2008). There are two reactions involved in the chemical process.In first reaction, generation of N_2 or N_3 takes place and is called homogeneous gas-phase reaction. In the second reaction, plasma interacts with other matters (solid or liquid) and is called heterogeneous reaction. In case of reactions involving solid surface, they are classified as: i)the onein which plasma-induced etching or ablation is used to remove material from the solid surface; ii) the another one in which concrete surface is deposited with a thin-film of material, and iii) the last one is marked with an addition or removal of any material. However, on exposure to plasma, there is physio-chemical alteration of the substrate surface (d'Agostino *et al.*, 2005). Parameters which are essential for the effective plasma generation and processing are:

- *Voltage:* Plasma generation needs breakdown voltage which is influenced by gas pressure and the distance between the two electrodes. Voltage increment directly influences the plasma generation; high voltage means high energy transfer into the plasma (Ziuzina *et al.*, 2013).
- *Pressure:* The available gas-particle density is low at low pressure, but the frequency of collision is high and energy acquired by electrons is not enough to ionize at high pressure (Bárdos and Baránková, 2010).
- *Type of gas:* Numerous noble gases, air or water vapour can be used in cold plasma. Out of 75 gases involved in 500 complex reactions (Gordillo-Vázquez, 2008), high oxygen gas is more effective for the plasma generation (Wan *et al.*, 2017; Misra *et al.*, 2014d).

- *Treatment time:* The reactive concentration of gas species and exposure time of the sample are directly influenced by treatment time (Ziuzina *et al*., 2013, 2014).

In order to generate plasma, suitable gas is placed between the two electrodes to generate electric fields with the constant or alternating current field. Various methods can achieve plasma state energy, including electric or magnetic field, thermal, microwave frequencies and many more. For food applications plasma generation is accomplished by employing corona discharge, plasma jets, dielectric barrier (DB) and gliding arc discharge; radiofrequency and microwave discharges (Misra, 2011; Surowsky *et al*., 2015). Some of the familiar sources are described below:

Corona discharge can be created in a modest device, where a sharp curve electrode is present. At a sharp electrode, the level of the electric field is substantial to produce ionization energy (Phan *et al*., 2017). Plasma jets operate in an open electrode system; the electrode placed outside is substantiated whereas, theother electrode gets polarizedby radio frequency (RF) power of 13.56 MHz. These electrodes are arranged in a coaxial manner. Plasma jets generates flames within the RF range at atmospheric pressure which facilitates the process of gas discharge (Nishime *et al*., 2017). The rigid collision of the free electron which got expedited by the RF field with background gases, produce various atoms and molecules that leaves the nozzle at elevated velocity (Misra *et al*., 2016).

A DB discharge has two electrodes whose layer covered up with quartz, glass or ceramic which keeps check on electric current leakage. The operating frequency for DBD is 0.05-500 kHz, the atmospheric pressure of 0.1-1 atm and the wide scope of gas pressures typically 10^4-10^6 Pa (Zhang *et al*., 2017).

Gliding arc discharge operates in the open air with 3mm thick electrodes where they are attached to knobs from both ends. The prerequisite of system is 60 Hz of alternating current (AC) and an output condition of <60 mA at 15kV (Niemira and Sites, 2008).

10.4 Prospective of Cold Plasma in Food Processing

10.4.1 Fortification

Micronutrient deficiency or micronutrient malnutrition also termed as hidden hunger is a serious global issue. Among various micronutrient deficiency Iron deficiency anaemia (IDA), which is referred to a condition in which blood does not have sufficient healthy red blood cells. IDA is the most common condition which affects approx. 20% of world's inhabitants (McLean *et al*.,

2009). India stands first in the prevalence of IDA with 39.86% population affected.Therefore, the issue of micronutrient deficiency must be addressed; food fortification can be a processing approach to overcome such problems.

Fortification of white rice with iron (Fe) and ascorbic acid was done using cold plasma technology by Akasapu *et al.* (2020). Treatment on white rice was performed at a fixed plasma voltage of 20KV for 10 min & 15 min. Thermal properties, % hydrophilicity and surface characteristics are detrimental factors for exposure time. Their outcomes were as follows: Significant increase in surface area verified by SEM images. The surface etching was found to be directly proportional to the treatment time. Reduction in cooking time by 10 minutes as compared to untreated rice. This phenomenon was attributed to increased surface energy and hydrophilicity of the rice. There was a significant enhancement of vitro bioavailability of Fe at pH 1.35 as well as at 7.5 for plasma-treated rice than of control.

10.4.2 Decontamination

Foodborne diseases are mainly caused by consuming foods that are infected by harmful microorganisms such as *Escherichia Coli,Listeria monocytogen, Enterococcus faecalis, Salmonella typhimurium* and *Staphylococcus aureus.* The effect of these microorganisms are not only limited to a disturbance in the gastrointestinal tract, but their severity can even lead to death in some cases (Yun *et al.*, 2010). Bacterial biofilms being a significant issue in the food industry causes significant losses to bothpublic and food industry. Biofilms on food dispensing equipment, food contact surfaces and filtered water supply system is linked with food spoilage, cross-contamination and foodborne disease outbreaks (Kim and Wei, 2012).

Although, there is a number of treatments generally used for sanitation and decontamination purpose such as the use of disinfectants (chlorination, hydrogen peroxide), chemical preservatives, mild heat treatments, microwave processing and ionizing radiation etc. Among various physical and chemical methods for cold decontamination, plasma has been proven to be highly efficient for diminishingtoxic contaminants in fresh foods (Niemira, 2012).

10.4.3 Modification of Starch

Starch being the most pervasive and broadlyutilized biological macromolecule, employed to food, textile, paper and pharmaceutical industries for technological purposes (Thirumdas *et al.*, 2017). Despite many advantages and a broad range of application in the food industry, starch when used in its native form exhibits several drawbacks, which narrows its application (Okyere *et al.*, 2019). The significant drawbacks of starch include the inability to withstand high temperature and shear conditions, insolubilities, unreactive nature, and retrogradation tendency. Therefore, to overcome such drawbacks, starch is generally exposed to the number of treatments (physical, chemical & enzymatic) to be modified in such a way as to meet the enormous market demand with improved technological and unique characteristics. Cold plasma technology is one of the physical methods of starch modification that has been recently used to modify the starch & eliminate the use of chemicals. Thirumades*et al.*(2017) studied the behaviour of starch after exposing it to cold plasma technology.They found that there was a significant decrease in molecular weight; gelatination temperature and viscosity. The alteration in such properties is because of depolymerization and cross- linking between amylose and amylopectin side chains.

Okyere *et al.*(2019) exposed three types of waxy starches, namely maize, potato and rice starch. They used CO2 and argon genes at flow rate of 25 standard cubic centimetres per minute and 15 standard cubic centimetres per minute to generate plasma. A radio frequency power of 120W and 0W was applied to the samples. Their findings revealed, increased gelatinization enthalpy for all 3 waxy starch varieties. There was also a significant increment in resistant starch content in all three waxy starches with increasing gas treatment. The decrease in crystallinity of potato starch was also observed but not in case of maize and rice starches.

10.4.4 Seed Germination

Cold plasma is a promising technology in agriculture, causing less harm to the crops, seeds, food and environment. Plasma technology contributes to the decontamination of seeds and plants, sowing, seed germination, storage, production of fertilizers (Los *et al.*, 2019; Grave *et al.*, 2019). Improvement in seed germination will not only be beneficial for the agricultural field but also in alcoholic beverages, weaning foods and speciality flours, where seed germination is the primary operation.Different methods were evaluated by researchers to enhance seed germination including magnetic field (Alpsoy and Unal, 2019; Hussain et al., 2020), microwave irradiation (Lazim and Ramadhan, 2020), ultrasound (Babaei Ghaghelestany *et al.*, 2020) and many more non-thermal technologies. Seeds, better yield and generation capacity can be accomplished by treating them with plasma as the penetrating capacity of plasma through the seed coat is higher; further, it straight away has an influence on the inside cell. The rate of germination got affected by ablation formed on the surface of seeds due to plasma, thus increasing the transmission of moisture and oxygen through the embryo (Thirumdas *et al.*, 2015). In a study, cold plasma with varied radiofrequency (RF) power (30, 90, 150, 210 and 270 W) were utilized by Singh *et al.* (2019) to evaluate the effect on sweet basil, showing that 150 W cold plasma treatment increases the germination, seedling vigor index I and II by 20, 48 and 66 per cent respectively. Atmospheric cold plasma treatment of 30-60 sec showed a significant surge in germination and seedling growth of wheat, increment in water uptake which may be due to surface changes caused by plasma treatment (Los *et al.*, 2019). Wide range of studies has been conducted on brown rice (Yodpitak *et al.*, 2019), pea seed (Gao *et al.*, 2019), fresh cuminseeds (Shashikanthalu *et al.*, 2020) and radish seed (Degutytė-Fomins *et al.*, 2020).

10.4.5 Milk and Milk Products

Milk and milk products are consumed worldwide and considered as highly nutritious and healthy diet. To inactivate pathogenic microbes and to reduce spoilage organisms in milk, needs thermal processing which ultimately affects the quality degrading fat, changes in protein structure and producing some secondary metabolites (aldehyde, ketone and acids) that can affect human health (Gavahian *et al.*, 2018). Therefore, plasma technology emergedas a viable technique forinactivation of harmful microorganisms, exposing them to positively charged reactive species that damages the outer layer of bacteria renders them inactive (Priyadarshini *et al.*, 2018). Coutinho *et al.* (2019) evaluated the outcome of low-pressure cold plasma technology with pasteurization on chocolate milk physico- chemical characteristics. The study revealed that different processing time of 5 to 15 minutes and the process stream rate of 10 to 30 mL/min does affect the characteristics. Higher inhibitory action of ACE and total phenolic content was observed when treated with high stream rate of gas (30 mL/min) while antioxidant activity remained unchanged compared to the product on whichpasteurization was employed. Increment in saturated fatty acid (SFA) content when treated with low and high level of processing conditions while a decrease in mono and poly-unsaturated fatty acid (UFA) since plasma induced oxygen radicals to react with UFA breaking their double bonds (Gavahian *et al.*, 2018). Cold plasma plays a significant role in disinfecting liquid milk as demonstrated by some recent studies (Wu *et al.*, 2020), whey based beverage (Silveira *et al.*, 2019), decontamination of cheese slices and production (Huang *et al.*, 2020; Wan *et al.*, 2019).

10.4.6 Fruit Juices

Cold plasma is very useful in processing products having high heat sensitivity. Fruit juices are lone that product which can mislay their appealability and functional value if processed thermally (Shi *et al.*, 2011). Almeida *et al.* (2015) investigated the role of ozone and plasma on prebiotic orange juice quality and observed that both processes degrade the oligosaccharide content partially,

Table 1: Fruit Juices' Processing by Cold Plasma

Type of Fruit Juice	Type of plasma employed	Variables	Modification in fruit juice	Reference
Siriguela	Indirect plasma	Gas- nitrogen Power supply-50 kHz Gas flow rate- 10-30ml/min Time- 5-15min Volume of sample-80 mL	• Increment in Pigments, vit- B total phenolics, and antioxidant activity • Vit-C and colourshowed no significant changes • Degradation in bioactive components with an increase in processing intensity of plasma	Paixão *et al*. (2019)
Blueberry	Single electrode atmospheric jet	Gas- Oxygen Gas concentration- 0, 0.5 and 1% Voltage- 11kV Frequency- 1000 Hz Time- 2,4 and 6 min	• Increase in time and O_2 concentration cause a decrease in Bacillus strain • Antioxidant activity increases with increasing O_2 concentration. • Better colour and phenolic content than thermal processing	Hou *et al*. (2019)
Cashew apple	In-direct plasma	Gas- Nitrogen Electric field- 80 kHz Gas stream rate- 10, 30 and 50 mL/min Treatment time- 5, 10 and 15 min Sample volume- 10 mL Vacuum- 30kPa	• Low gas flow rate promotes Vit-C, phenolic content and antioxidant activity • Higher flow rate helpful to increase flavonoids and polyphenols • Overexposure negatively affected the bioactive components	Rodríguez *et al.* (2017)
Apple	DBD plasma	Gas- air Power- 30, 40 and 50 W Treatment time- 0 to 40 sec Sample volume- 3 mL	• The cell membrane of E.Coligot damaged with treatment • Max. power for more than 10 sec results in a significant change in pH, TA, colour and total phenolic content	Liao *et al.* (2018)

whereas, treatment time of 15-60 sec showed no substantialvariations in total phenolic content and antioxidant capacity of orange juice. Summarization of the effect of cold plasma on the quality and functionality of various fruit juices are mentioned in Table 1.

10.4.7 Phenolic and Antioxidant Compounds

Consumers' attraction towards minimally processed products, their preservation and quality maintenance in terms of their biological activity have fostered the development of cold plasma technology. The various health-promoting effects provided by phenolics and antioxidants compounds are the growing interest in the food sector. The ability of antioxidants to defend against free radical species and their interaction with plasma is the area of interest for many researchers. Ramazzina *et al.* (2016) explored the DBD plasma generator effect on the functional property of fresh-cut apples.Plasma cause10% reduction in antioxidant capacity though it did not affect the cell response to oxidative stress. Freshly cut pears treated with water activated with plasma showed a slight positive difference in radical scavenging activity in the starting days of the storage, however, after storage of eight days, no significant difference was observed (Chen *et al.*, 2019). Atmospheric pressure plasma jet was employed by Grzegorzewski *et al.* (2011) to study lamb's lettuce phenolic profile. The outcomes showed that phenolic acid levels were reduced whereas diosmetin level increases.

Further, there was no change observed in chlorogenic acid content. Bao *et al.* (2020) conducted a study for improving phenolic extractability from tomato pomace, discloses 10% increment in phenolic extract yield when treated with He and N2 gases. Fresh-cut pitaya fruit was subjected to cold plasma treatment at 60kV for 5 minute, outcomes revealed that there were significant enhancement of phenolic accumulation, antioxidant capacity and acceleration of rate of generated ROS (reactive oxygen species) than that of control (Li *et al.*, 2019).

10.4.8 Food Packaging Material

Food packaging material offers protection to the food product from the outside environment throughout the food supply chain. Apart from disinfecting packaging material, cold plasma also has its application in developing edible packaging material. Cold plasma can help in surface modification of packaging material, pretreatments of the surface like cleaning, passivation, printing and adhesion (Pankaj *et al.*, 2014). Several packaging materials can be sterilized quickly and without upsetting the material's composition or leaving any residues, via low-temperature gas plasma sterilization. Packing materials including polyethylene and polycarbonate which are heat-sensitive can be sterilized at low temperature via the use of cold plasma. In particular, films that are edible, hydrophobicity nature of the polymer surface with low energy is desirable for sterilization (Vesel and Mozetic, 2012). Cold plasma treatment given to polylactic acid films (Song *et al.*, 2016) showed that the treatment improved printability and biodegradability of the films. Cui et al. (2020) utilized *Momordica charantia* polysaccharide (MCP) which served as nanofiber matrix for encapsulating phlorotannin (PT). This nanofiber membrane when exposed to cold plasma, the releasing efficiency of PT was heightened by 23.4% (4°C) and 25% (25°C).

Fish myofibrillar proteins as a bi-layer film was researched by Romani *et al.* (2020). They treated the film with glow discharge plasma and coated with wax (carnauba), their findings are:high tensile strength of 175% and low water vapour permeability of 65% in comparison to control. Cold plasma strategy improved biopolymer film properties specially fishmyofibrillar protein films. Cold plasma for period of 2 min resulted in increased elongation and water vapour permeability, whereas, 5 min treatment time decreases the tensile strength and increased water solubility (Romani *et al.*, 2019). Suitability of any packaging material for food applications is greatly influenced by its barrier properties, therefore, there has been a continuous effort to accomplish enhanced barrier properties through application of cold plasma (Thirumdas *et al.*, 2015).

10.5 Limitations of Cold Plasma

Cold plasma presents potential opportunities in the field of non-thermal technologies for the food industry. Along with the numerous advantages of processing food with cold plasma, we also wish to mention few limitations of plasma processing like diminishing colour intensity and firmness of fruits, increment in lipid oxidation and acidity. Food structure should be the primary focus for designing plasma treatments ensuring food safety and efficient processing of food. Some of the internal factors of food such as pH and osmolarity can affect the plasma treatments resulting in hardening of bacteria making them more resistant, whereas, presence of lipid, proteins and bioactive compounds can sincerely diminish the efficiency of plasma species (Smet *et al.*, 2016).

Plasma treated food products like fresh pork showed lighter colour due to the possible reaction involved between myoglobin protein and hydrogen peroxide produced during plasma treatment (Fröhling *et al.*, 2012) along with some other fresh produce, tomatoes, green lettuce and carrots (Bermúdez-Aguirre *et al.*, 2013). Foods that contain a high amount of fat can form reactive oxygen species which may further result in lipid oxidation on reacting with gas mixture contained in cold plasma. A significant complication associated with cold plasma treatment of foods is the hindrance in the ability to regulate the gas plasma reactions' chemistry which stems from introduction of fluctuating humidity by foods which. Further, to carry out plasma processing, very high voltages are required which requires additional safety measures. For the cold plasma's applicability in the food industry, economic cost of the gas, the mixture of gases or noble gases should be assessed. As discussed earlier, the advantages of cold plasma to produce high-quality products could outweigh its higher cost (Li and Farid, 2016).

10.6 Future Panoramas and Conclusion

Cold plasma is the new upcoming technology in the non-thermal world which is safe for the environment, requires low temperature to inactivate microorganisms, fulfils all ecological standards and maintains the product quality. All new technologies face some challenges in food industries as the competitive pressure builds up on small and medium-sized enterprises, whether, it is the cost of the equipment, the involvement of risk, the effect on the end product nutritional and quality characteristics. Cold plasma technology to be successfully adopted in the food industry, needs to satisfy not only food safety factor but also the consumer. Consumers are the primary focus when considering any novel technology in the production line. Nowadays, what drives consumers towards food are quality and nutrition, minimal use of preservatives and other chemicals. Additionally, there is also a young subset of consumers that consider environmental sustainability, the main factor in food production. Apart from consumers, factors which interest producers, processors, distributors and retailers are: potential increment in product shelf life, lower consumer and processing wastage or loss, better retention of food product quality and nutritional factor, low requirement of energy, operational and maintenance cost.

Widespread use of plasma technology needs regulatory approval; every country has its own regulatory standards for food process interventions that proved to be an obstacle. Plasma chemistry is very complex, which needs to be explored, prioritized and assessed. Active involvement of plasma community is must to regulate review process for fast commercialization of plasma technology. Proper guiding documents or a roadmap is required describing plasma generation and its types, practices employed for data collection, measurement and analysis of plasma.

Being the cold processing technology, it is conducive in preserving the product quality characteristics. In this respect, it can be concluded that in the near future, we expect that cold plasma becomes standard processing technology in the food sector.

10.7 References

Akasapu K, Ojah N, Gupta AK, Choudhury AJ and Mishra P. 2020. An innovative approach for iron fortification of rice using cold plasma. Food Res. Int, 136, 109599, 2020.

Alpsoy HC and Unal H. 2019. Effect of stationary magnetic field on germination and crop yield in spinach (*Spinacia oleracea* L.). Comptesrendus de l'Académiebulgare des Sci, 72(5), .

Almeida FDL, Cavalcante RS, Cullen PJ, Frias JM, Bourke P, Fernandes FA and Rodrigues S. 2015. Effects of atmospheric cold plasma and ozone on prebiotic orange juice. Inn. Food Sci. Emer. Technol, 32, 127-135.

Babaei-Ghaghelestany A, Alebrahim MT, MacGregor DR, Khatami SA and Hasani Nasab Farzaneh, R. 2020. Evaluation of ultrasound technology to break seed dormancy of common lambsquarters (*Chenopodium album*). Food Sci. Nutri.

Bao Y, Reddivari L and Huang JY. 2020. Development of cold plasma pretreatment for improving phenolics extractability from tomato pomace. Inn. Food Sci. Emer. Technol, 65, 102445.

Bárdos L and Baránková H. 2010. Cold atmospheric plasma: sources, processes, and applications. Thin Solid Films, 518 (23), 6705-6713.

Bermúdez-Aguirre D, Wemlinger E, Pedrow P, Barbosa-Cánovas G and Garcia-Perez M. 2013. Effect of atmospheric pressure cold plasma (APCP) on the inactivation of Escherichia coli in fresh produce. Food Control, 34(1), 149-157.

Chen C, Liu C, Jiang A, Guan Q, Sun X, Liu S and Hu W. 2019. The effects of cold plasma-activated water treatment on the microbial growth and antioxidant properties of fresh-cut pears. Food Bioprocess Technol, 12(11), 1842-1851.

Coutinho NM, Silveira MR, Fernandes LM, Moraes J, Pimentel TC, Freitas MQ, and Neto RP. 2019. Processing chocolate milk drink by low-pressure cold plasma technology. Food chem, 278, 276-283.

Cui H, Yang X, Abdel-Samie MA and Lin L. 2020. Cold plasma treated phlorotannin/ Momordica charantia polysaccharide nanofiber for active food packaging. Carbohydrate Polym, 116214.

d'Agostino R, Favia P, Oehr C and Wertheimer MR. 2005. Low temperature plasma processing of materials: Past, present, and future. Plasma Processes Polym, 2(1), 7-15.

Degutytė-Fomins L, Paužaitė G, Žūkienė R, Mildažienė V, Koga K and Shiratani M. 2020. Relationship between cold plasma treatment-induced changes in radish seed germination and phytohormone balance. Japanese J. Appl. Phys, 59(SH), SH1001, 2020.

Ekezie FGC, Sun DW and Cheng JH. 2017. A review on recent advances in cold plasma technology for the food industry: Current applications and future trends. Trends Food Sci. Technol, 69, 46-58.

Fridman A. 2008. *Plasma chemistry*, Cambridge Univ Press, New York.

Fröhling A, Durek J, Schnabel U, Ehlbeck J, Bolling J and Schlüter O. 2012. Indirect plasma treatment of fresh pork: Decontamination efficiency and effects on quality attributes. Innov. Food Sci. Emerg. Technol, 16, 381-390.

Gao X, Zhang A, Héroux P, Sand W, Sun Z, Zhan J, and Guo Y. 2019. Effect of dielectric barrier discharge cold plasma on pea seed growth. J. Agricul. Food Chem, 67(39), 10813-10822, 2019.

Gavahian M, Chu YH, Khaneghah AM, Barba FJ and Misra NN. 2008. A critical analysis of the cold plasma induced lipid oxidation in foods. Trends Food Sci. Technol, 77, 32–4 .

Gordillo-Vázquez FJ. 2008. Air plasma kinetics under the influence of sprites. J. Phys. D: Appl. Phys, 41 (23), 234016.

Graves DB, Bakken LB, Jensen MB and Ingels R. 2009. Plasma activated organic fertilizer. Plasma Chemis. Plasma Processing, 39(1), 1-19.

Grzegorzewski F, Ehlbeck J, Schlüter O, Kroh LW and Rohn S. 2011. Treating lamb's lettuce with a cold plasma–Influence of atmospheric pressure Ar plasma immanent species on the phenolic profile of Valerianellalocusta. LWT-Food Sci. Technol, 44(10), 2285-2289.

Hou Y, Wang R, Gan Z, Shao T, Zhang X, He M and Sun A. 2019. Effect of cold plasma on blueberry juice quality. Food chem, 290, 79-86.

Huang YM, Chen CK and Hsu CL. 2020. Non-thermal atmospheric gas plasma for decontamination of sliced cheese and changes in quality. Food Sci. Technol. Int, 1082013220925931.

Hussain MS, Dastgeer G, Afzal AM, Hussain S and Kanwar RR. 2020. Eco-friendly magnetic field treatment to enhance wheat yield and seed germination growth. Environ. Nanotechnol, Monitoring Manag, 100299.

Kim SH and Wei CI. Biofilms. 2012. Decontamination of fresh and minimally processed produce, 59-75.

Lazim SK and Ramadhan M. 2020. Study effect of a static magnetic field and microwave irradiation on wheat seed germination using different curves fitting model. J. Green Eng, 10, 3188-3205.

Li X and Farid M. 2016. A review on recent development in non-conventional food sterilization technologies. J. Food Eng, 182, 33-45.

Li X, Li M, Ji N, Jin P, Zhang J, Zheng Y and Li F. 2019. Cold plasma treatment induces phenolic accumulation and enhances antioxidant activity in fresh-cut pitaya (Hylocereusundatus) fruit. LWT, 115, 108447.

Liao X, Li J, Muhammad AI, Suo Y, Chen S, Ye X and Ding T. 2018. Application of a dielectric barrier discharge atmospheric cold plasma (Dbd-Acp) for Eshcerichia coli inactivation in apple juice. J. Food Sci, 83(2), 401-408.

Los A, Ziuzina D, Boehm D, Cullen PJ and Bourke P. 2019. Investigation of mechanisms involved in germination enhancement of wheat (Triticum aestivum) by cold plasma: Effects on seed surface chemistry and characteristics. Plasma Processes Polym, 16(4), 1800148.

Man, P. 2005. Microbial inactivation by new technologies of food preservation. 1387–1399, 2005.

McLean E, Cogswell M, Egli I, Wojdyla D and De Benoist B. 2009. World widep revalence of anaemia, WHO vitamin and mineral nutrition information system, 1993–2005. Public HealthNutrition, 12(4), 444-454.

Misra NN, Moiseev T, Patil S, Pankaj SK, Bourke P and Mosnier JP. 2014d. Cold plasma in modified atmospheres for post-harvest treatment of strawberries. Food Bioprocess Technol, 7 (10), 3045-3054.

Misra NN, Pankaj SK, Segat A and Ishikawa K. 2016. Cold plasma interactions with enzymes in foods and model systems. Trends Food Sci. Technol, 55, 39-47.

Misra NN, Schlüter O and Cullen PJ. (Eds.), 2016. Cold plasma in food and agriculture: fundamentals and applications. Academic Press.

Misra NN, Tiwari BK, Raghavarao KS and Cullen PJ. 2011. Nonthermal plasma inactivation of food-borne pathogens. Food Eng. Reviews, *3*(3-4), 159-170.

Niemira BA and Sites J, 2008. Cold plasma inactivates Salmonella Stanley and Escherichia coli O157:H7 inoculated on golden delicious apples. J. Food Protection, 71(7), 1357-1365.

Niemira BA. 2012. Cold plasma decontamination of foods. Annual review food sci technol, 3, 125- 142.

Nishime TMC, Borges AC, Koga-Ito CY, Machida M, Hein LRO and Kostov KG. 2017. Non- thermal atmospheric pressure plasma jet applied to inactivation of different microorganisms. Surface Coatings Technol, 312, 19-24.

Okyere AY, Bertoft E and Annor GA. SC. 2019. Carbohydrate Polym, 115075.

Paixão LM, Fonteles TV, Oliveira VS, Fernandes FA and Rodrigues S. 2019. Cold plasma effects on functional compounds of siriguela juice. Food Bioprocess Technol, 12(1), 110-121.

Pankaj SK, Bueno-Ferrer C, Misra NN, Milosavljević V, O'donnell CP, Bourke P, Keener KM and Cullen PJ. 2014. Applications of cold plasma technology in food packaging. Trends Food Sci. Technol, 35(1), 5-17.

Phan KTK, Phan HT, Brennan CS and Phimolsiripol Y. 2017. Nonthermal plasma for pesticide and microbial elimination on fruits and vegetables: an overview. Int. J. Food Sci Technol. 52, 2127–2137, 2017.

Priyadarshini A, Rajauria G, O'Donnell CP and Tiwari BK. 2019. Emerging food processing technologies and factors impacting their industrial adoption. Critical Reviews Food Sci. Nutrition, 59(19), 3082-3101.

Ramazzina I, Tappi S, Rocculi P, Sacchetti G, Berardinelli A, Marseglia A and Rizzi F. 2016. Effect of cold plasma treatment on the functional properties of fresh-cut apples. J. Agricul. Food Chem, 64(42), 8010-8018.

Rodríguez Ó, Gomes WF, Rodrigues S and Fernandes FA. 2017. Effect of indirect cold plasma treatment on cashew apple juice (Anacardiumoccidentale L.). LWT, 84, 457-463.

Romani VP, Olsen B, Collares MP, Oliveira JRM, Prentice C and Martins VG. 2020. Cold plasma and carnauba wax as strategies to produce improved bi-layer films for sustainable food packaging. Food Hydrocoll, 10608.

Romani VP, Olsen B, Collares MP, Oliveira JRM, Prentice-Hernández C and Martins VG. 2019. Improvement of fish protein films properties for food packaging through glow discharge plasma application. Food hydrocoll, 87, 970-976.

Shashikanthalu SP, Ramireddy L and Radhakrishnan M. 2020. Stimulation of the germination and seedling growth of Cuminumcyminum L. seeds by cold plasma. J. Appl. Res. Medicinal Aromatic Plants, 100259.

Shi XM, Zhang GJ, Wu XL, Li YX, Ma Y, Shao XJ. 2011. Effect of low-temperature plasma on microorganism inactivation and quality of freshly squeezed orange juice. Plasma Sci., IEEE Trans. 39, (7), 1591-1597, 2011.

Siddique SS, Hardy GSJ and Bayliss KL. 2018. Cold plasma: a potential new method to manage postharvest diseases caused by fungal plant pathogens. Plant Pathology, 67(5), 1011-1021.

Silveira MR, Coutinho NM, Esmerino EA, Moraes J, Fernandes LM, Pimentel TC, and Borges FO. 2019. Guava-flavored whey beverage processed by cold plasma technology: Bioactive compounds, fatty acid profile and volatile compounds. Food chemistry, 279, 120-127, 2019.

Smet C, Noriega E, Rosier F, Walsh JL, Valdramidis VP and Van Impe JF. 2016. Influence of food intrinsic factors on the inactivation efficacy of cold atmospheric plasma: Impact of osmotic stress, suboptimal pH and food structure. Innov. Food Sci. Emerg. Technol, 38, 393-406.

Singh R, Prasad P, Mohan R, Verma MK and Kumar B. 2019. Radiofrequency cold plasma treatment enhances seed germination and seedling growth in variety CIM-Saumya of sweet basil (Ocimumbasilicum L.). J. App. Res. Med. Aromatic Plants, 12, 78-81.

Song AY, Oh YA, Roh SH, Kim JH and Min SC. 2016. Cold oxygen plasma treatments for the improvement of the physicochemical and biodegradable properties of polylactic acid films for food packaging. J. Food Sci, 81(1), E86-E96.

Surowsky B, Schlüter O and Knorr D. 2015.Interactions of Non-Thermal Atmospheric Pressure Plasma with Solid and Liquid Food Systems: A Review. Food Eng. Reviews, 7, 82-108.

Thirumdas R, Kadam D and Annapure US. 2017. Cold plasma: an alternative technology for the starch modification. Food Biophysics, 12(1), 129-139.

Thirumdas R, Sarangapani C and Annapure US. 2015. Cold plasma: a novel non-thermal technology for food processing. Food biophysics, 10(1), 1-11.

Vesel A and Mozetic M. 2012. Surface modification and ageing of PMMA polymer by oxygen plasma treatment. Vacuum, 86(6), 634-637.

Wan Z, Chen Y, Pankaj SK and Keener KM. 2017. High voltage atmospheric cold plasma treatment of refrigerated chicken eggs for control of Salmonella Enteritidis contamination on egg shell. LWT-Food Sci. Technol, 1(76), 124-30.

Wan Z, Pankaj SK, Mosher C and Keener KM. 2019. Effect of high voltage atmospheric cold plasma on inactivation of Listeria innocua on Queso Fresco cheese, cheese model and tryptic soy agar. LWT, 102, 268-275.

Wu X, Luo Y, Zhao F and Mu G. 2020. Influence of dielectric barrier discharge cold plasma on physicochemical property of milk for sterilization. Plasma Proc. Polym, e1900219.

Yodpitak S, Mahatheeranont S, Boonyawan D, Sookwong P, Roytrakul S and Norkaew O. 2019. Cold plasma treatment to improve germination and enhance the bioactive phytochemical content of germinated brown rice. Food chemistry, 289, 328-339.

Yun H, Kim B, Jung S, Kruk ZA, Bee D, Choe W and Jo C. 2010. Inactivation of Listeria monocytogenes inoculated on disposable plastic tray, aluminum foil, and paper cup by atmospheric pressure plasma. Food Control, 21(8), 1182–1186.

Zhang H, Ma D, Qiu R, Tang Y and Du C. 2017. Non-thermal plasma technology for organically contaminated soil remediation: A review. Chem. Eng. J, 313, 157-170.

Ziuzina D, Patil S, Cullen PJ, Keener KM and Bourke P. 2013. Atmospheric cold plasma inactivation of Escherichia coli in liquid media inside a sealed package. [Research Support, Non-U.S. Gov't]. J. Appl. Microbiol, 114 (3), 778-787.

Ziuzina D, Patil S, Cullen PJ, Keener KM and Bourke P. 2014. Atmospheric cold plasma inactivation of Escherichia coli, Salmonella enterica serovar Typhimurium and Listeria monocytogenes inoculated on fresh produce. Food Microbiol, 42, 109-116.

11

Cooked Food Waste Management An Eco-friendly Approach

K. Sai Shiva, Anurag Singh, Rakhi Singh and S. Thangalakshmi[2]*

ABSTRACT

The improper disposal of food waste poses an environmental threat. The management of food waste in a sustainable way is always a challenge. Food waste is generated at various stages starting from food production till consumption. The cooked food waste generated due to various reasons leads not only to a loss of the food produced but also the water and energy used to cook it. This cooked food leftover can be served to the needy people or if not possible, can be processed further. This chapter focuses on various technological options for the management of cooked food waste in a sustainable way.

Keywords: Cooked food waste, Compost, Bokashi, Bio-surfactants, Environment

11.1 Introduction

Food waste is an ecological, economic and social problem. Food wastage means the loss of any food item either by waste or deterioration.Thus the term – wastage encompasses both food loss and food waste. Nearly one third of global food produced for human consumption is getting waste or lost which is equal to 1.3 billion tons of food per year (FAO, 2018). Normally this is due to food spoilage but factors such as excess supply to markets, change in consumer eating/shopping habits also contribute to the problem.Wastage of food in India is a serious issue. It is accounted that every year around 70 million tons of food is getting wasted which is having a value of Rs. 93000 crore. Studies say that the food wasted in India is equal to what the whole U.K. consumes (Kulshrestha and Richa, 2017). On the other hand according to 2019 global hunger index

*Corresponding Author

(GHI) report, India ranks at a position of 102 among 117 countries in terms of severity of hungry where as India's GHI score was enormously increased from 17.8 in 2014 to 30.3 in 2019 (PHD Research Bureau, 2020). About 25% of the hunger people worldwide lives in India. Indian government took several steps in managing food waste such as running food collection vehicles & centers to feed the needy, establishing campaigns to educate the people regarding food waste etc (Abu-Khader, 2006). But still a number of children in India are malnourished. The main reason for this malnourishment is the unavailability of the food to the children. Un-availability doesn't mean that India is not producing the enough food but it is the wastage of food that is occurring. In this situation, we cannot afford wasting such a huge amount of food. Food wastage happens at different stages after the harvesting.

More serious issue is the wastage of cooked food or leftovers. The picture is very frightening globally. We can easily see the huge amount of cooked food being thrown away in urban areas whereas the people in rural areas or deprived areas are not getting two time meals (Kulshrestha and Richa, 2017). Major portion of this food waste turns up in landfills from which methane will emit, a major greenhouse gas. This cooked food wastage is not only a wastage of nutrients but of water used for irrigation, energy, man power involved in cultivation and preparation, and electricity too(Koppel *et al.*, 2016). The cooked food wastage also affects the environment adversely. Without a proper planning for such solid waste management, this will create a threat to environment and human health (Kulshrestha and Richa, 2017).

11.2. Reasons for Cooked Food Waste

Generally food waste occurs in any stage of supply chain, it may lies in any of the hands of producers, processors, retailers, consumers individually or combination of more than one. But the major cooked food waste is prone to be from households of higher socio-economic groups, restaurants, function halls etc. The major reason for this cooked food wastage can be as follows:

11.2.1 Lack of Appropriate Planning

Food waste is mainly due to lack of proper planning in households, restaurants or any functions etc.In several cases the cooked food may lead to expiry due tobusy schedulesof work and appointments which lead to change in food preparation plan or fail to use it on time. Improper training lead people like themselves badly and preparations which are not having good resulting value (US EPA, 2020).

11.2.2 Over Preparation of Food in Hotels, Restaurants and Food Catering Services etc.

This type of services generally lack in quantifying amount of food consumed average and they wish to keep on producing enough but most of it is actually not needed. It was estimated that half of food waste generated from total solid waste which is coming from hotel industry and also from non compliances to the food waste strategies (Kasavan *et al.*, 2017). Their major intention in anticipation of high customer volume depends on ability of not running of menu which made them to over preparation of food that leads to wastage if food is unsold.In general, management directs to produce food in large batches in order to minimize the cost but in actual reality it results in more food waste in comparison to cook to order or small batch cooking. But other reasons such as the time they spent for implementation of waste management strategies and inefficiency of the team commitment towards managing their waste is also responsible for food waste (Mohan *et al.*, 2017).

11.2.3 Consumer Behavior

People wishes to have the food more than the requirement rather than which is exactly sufficient. Consumer's tendency on preferences forces food handlers in restaurants, hotels to have big menu and optimum levels of service with proper consistency which tends to large waste of food. Customer likes to taste cooked food items which is new to them in big hotels which leads to waste if they don't like it. Managing wish to avoid negative emotions like guilt, frustrate, regret will remove the barriers on minimizing food waste (Graham-Rowe *et al.*, 2014).

11.2.4 Large Menu in Social Gatherings

Now a day we can see that in each gathering we have a huge list of food items starting from starters, soups, main course, desserts etc. the guests hardly eat everything out of these. That also causes a huge wastage of cooked food (Bulluck Iii *et al.*, 2002).

11.2.5 Ineffective Utilization of the Cooked Food Waste

The best strategy for the utilization of the cooked food/ leftover is to distribute the food to the needy ones to avoid wastage. The untouched leftover food after any function may be distributed to the underprivileged hungry people by a number of NGOs those are working in this field (Kulshrestha and Richa, 2017). They come and collect the leftover food after a large scale function and distribute it among the needy people in orphanage, schools, old age homes, etc. So failure to implement this strategy also results in food waste.

11.2.5.1 Challenges in Distribution of Leftover Food

The food that has been cooked and is not yet served, starts losing its temperature. This milder temperature is suitable for microbial growth in long run. If such food is not handled carefully and distributed, may cause serious issue like food borne illness. This is why some of the caterers have their reservation in donating the leftover foods. Keeping the food safe till it is consumed is the biggest challenge. The cooked food that is stored at room temperature not at refrigerated temperature spoils very fast due to microbial growth (Kulshrestha and Richa, 2017). The time duration may vary based on the composition of the cooked food such as pH, moisture content, addition of spices etc. so it is required to maintain a refrigerated condition of food or any other suitable atmospheric condition (Kulshrestha and Richa, 2017).

11.3 3-R Strategy for Utilization of Cooked Food Leftover

This strategy is meant to find a solution for the cooked food wastage which expresses 3R as Reduce, Reuse and Recycle. Each of them has its own importance in the day to day life in managing cooked food loss and waste.

11.3.1 Reduce

Every house hold has to plan before cooking of food regarding how much and how to cook food. These two aspects are critical for controlling food loss and waste. Advanced planning of the meals is the best approach to controlexcessive buying of items to reduce buying items which are not needed and used(US EPA, 2020). If one can re-buy or re-purchase all perishables and fresh produces every few days than purchasing it on a bulk for a week's worth in the hope that makes use everything. Any food servicing operators who ever trying to prepare novel foods has to cook in very less amount because if it tastes bad then it goes waste. Technology can also play a role in controlling food waste by handling leftovers and which eventually increases profits and a new customer base development by positive impact of the establishment (Martin-Rios *et al*., 2018). In the present scenario availability of instant, ready to cook products are increasing day by day, usage of such type of products at the time of need definitely controls food loss. Restaurants or big hotels have to cook on the basis of daily average consumption and they have to opt for the instant cooking methods if the customer's volume increases suddenly. Presently the restaurants which are managing waste are few or low and the chefs who are creating meals out of waste are also very few(Martin-Rios *et al*., 2018).

11.3.2 Reuse

An approach that not necessarily falls within the households, restaurants or any food servicing operators is the reuse of food material as food. There are several methods which are derived for reusing the cooking food (Mohan *et al.*, 2017). The first and foremost point is refrigerating the cooked food within two hours and can be reused by the application of heat to it. Leftovers can be used not only by storing and reusing but they can also be converted into any form of food such as snacks by applying specific cooking methods. Actions to reuse the wastage of unsold food material such as food banks, emergency shelters have been initiated by government and non-government organizations.

11.3.3 Recycle

No industry wishes to produce processed food from collecting leftover cooked food from households because it creates huge losses. Instead they opt to run food collection vehicles and feed the needy. The concept of recycle arises if the cooked food is not fit for consumption. If the food which has recycled is used as an input like biomass or fertilizer, animal feed to give output then we can say that food is not getting waste. There is a possibility of incurring economic losses if the recovered food cost is more than average input cost in the non food use. To solve this issue, there are two methods which can be used in household level and industrially to recycle the cooked food waste. They are: (a)conversion of cooked food waste to homemade bio-compost by Bokashi process and (b) Conversion of cooked food to sophorolipids (bio-surfactants) industrially. These two processes reveal that food which is not fit for consumption can also boost the plant life and create benefits to the earth in eco-friendly way.

11.4. Conversion of Cooked Food Waste to Homemade Bio-Compost By Bokashi Process

The process which utilizes the concept of effective microorganisms to convert cooked food waste into an effective biofertilizer is known as —Bokashi process. In Japanese language Bokashi means to ferment.

Problems of root diseases and diminished soil health arise with the usage of synthetic fertilizers and pesticides in conventional agriculture (Bulluck Iii *et al.*, 2002). Soil borne diseases will lead to decrease yields in variety of crops. Although organic farming is evolved as an alternative to conventional agriculture methods but several pesticides of chemical origin are not allowed to use in it. Even the microbial inoculants like fungi or bacteria are produced like control agents of biological nature or fertilizers to manage plant pathogens, provide nutrients to crops, for plant growth and soil structure improvement but lacks in maintaining consistent and reproducible growth(Shin *et al.*, 2017). Then the concept of usage of beneficial and effective microorganisms of mixed cultures is evolved and effectively worked in improving soil health and quality, inevitably raising the yield and quality of crops (Xiaohou *et al.*, 2008).

Generally traditional composting method which we use in our home is cheaper to use but it doesn't allow putting cooked food waste, dairy products, meat and fishing as they produce bad odours which attracts rats and flies. But the Bokashi system of composting allows the cooked food waste and ends up nourishing the garden rather than landfill. The Bokashi process is depicted in Fig.1.

Fig. 1: Bokashi method (*Source*: Ambong *et al.*, 2018)

11.4.1 Working Process

Bokashi bins which specifically made for this process and used to store cooked food waste or leftovers of daily households made of a fitting lid tightly on top along with a spigot on bottom(Fig-2). Every day cooked food waste and kitchen waste has to be put inside the Bokashi bin(typically of 5-10kg),in case of meat bones are chopped to small pieces to increase its efficiency. This food waste is pressed and a layer of 3-4 cm of Bokashi bran which consists of effective microorganisms is spread on it (Bokashi, 2010). Bokashi bran consists of primarily different strains of yeasts (*Saccharomyces* spp.), bacteria which give lactic acids (*Lactobacillus spp.)*etc. (Bokashi Living, 2016). After adding these microbial inoculants to the food waste, the bin has to be closed air tight with lid which facilitates to undergo anaerobic fermentation of organic matter. In this way every day we have to add the cooked food waste to the bin and sprinkle the handful of Bokashi bran on them.

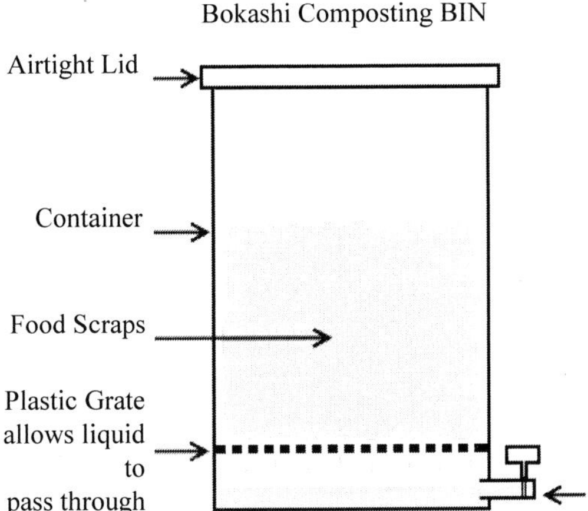

Fig. 2: Bokashi decomposition bin

To avoid foul smelling while opening the lid, it's better to store the food for one week and add it to the Bokashi bin once a week by spreading a layer of Bokashi bran on them. When the bin becomes full, it should be covered tightly and kept away from sunlight. When the fermentation of organic matter proceeds by the *Lactobacilli,* it starts to break down of carbohydrates into lactic acids by homolactic fermentation and secretes a brown liquid which is known as Bokashi tea which is to be collected at the bottom of bin by using tap (Bokashi, 2010). Bokashi tea consists of good amounts of proteins, nutrients and lactic acid.As the acid content in the tea is very high, upon dilution it can apparently acts as a fertilizer. Without dilution it can control slime in drains,pipes and septic-system.

Cooked food waste which is anaerobically fermented for ten to fourteen days is generally placed on soil matter which is ready to use or can be stored unopened for future use. Upon mixing in the soil, it interacts with air where pyruvate, a fundamental energy provider, is formed with oxidation of lactic acid which is soon consumed by soil (Bokashi, 2010). This technology can make farmers to adopt non chemical organic farming system with no environmental risks (Xiaohou *et al.*, 2008).

Anyone can enliven their garden by using this economical bio-compost and nutritious Bokashi tea. Households can maintain small orchards or kitchen garden by using this process. Like other forms of composting, Bokashi doesn't require complicated and lengthy process like aeration, exact amount of moisture, mix of green and brown etc. This process is affordable and increases the soil structure in retaining more water and reducing plant diseases.

11.5 Conversion of Cooked Food Waste to Sophorolipids (Bio-Surfactants) Industrially

Cooked food waste is not only used to make bio-compost or landfills, rather it can be used in the production of Sophorolipids (SLs). Generally sophorolipids consist of hydrophilic carbohydrates and hydrophobic fatty acids and they are described about 60 years ago(Gorin *et al.*, 1961). Sophorolipids belongs to glycolipid class of microbial surfactants produced by bio-coversion of hydrophobic substrates and hydrophilic saccharides by relatively high yields of several non-pathogenic yeast species and are secreted extracellularly under suitable production conditions. Mixed food waste which is generally collected from restaurants can be used as the most part of that waste would be rich in fatty acids and carbohydrates. The process is shown in fig.-3

Sophorolipids, the major market share holders of microbial bio-surfactants which has 3% global share in the surfactant market. Surfactants are agents which reduce the surface tension of liquid when it is added and they are used as detergents, wetting agents, emulsifiers, foaming agents etc. Surfactants market wasestimated as USD 30.64 billion in 2016 and is projected to USD 39.86 billion by 2021, with a CAGR of 5.4 % between 2016 and 2021(Global Forecast, 2020).

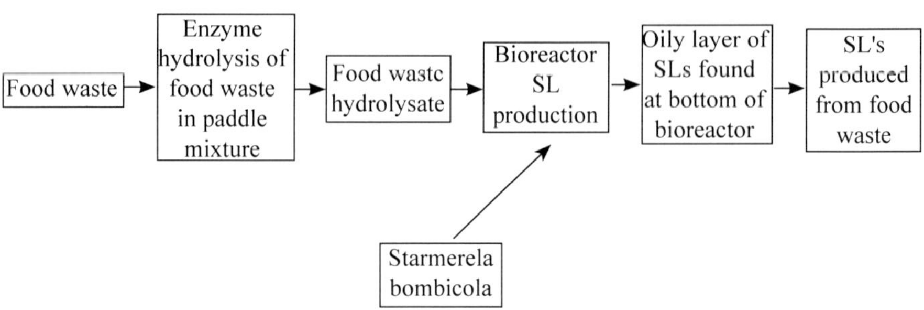

Fig. 3: Production process of sophorolipids

11.5.1 Production Process

The production process of sophorolipids mainly involves few stages like enzymatic hydrolysis, purification and fermentation. From all the restaurants, mixed food waste has to be collected which consists of noodles, rice, chicken fat, along with bones and vegetables (Kaur *et al.*, 2019). Bones and vegetables are to be separated before starting of the process.

11.5.1.1 Enzymatic Hydrolysis

Few enzymes are involved in the process of enzymatic hydrolysis. They are Protease (produced from *Bacillus subtilus*), lipase(produced from *Aspergillus niger*), glucoamylase (produced from *Aspergillus niger*). The enzyme glucoamylase is loaded at a ratio of 1% (v/w) whereas lipase and protease loaded at 1%(w/v) which is to be supplied in powder form. Hydrolysis should be carried in paddle mixer with agitation speed of 150 rpm at a temperature of 48°C and NaOH is used to adjust pH at 5.55 (Kaur *et al.*, 2019).

11.5.1.2 Purification Process

Hydrolysate has to be filtered before using it as feed stock in fermentation process. Due to gravity, hydrolysis mixture separates lipids (also used for sophorolipid production) at the top layer in the separating funnel leaving the hydolysate at the bottom. The hydrolysate which is used in the fermentation process has to be filtered to obtain sterilized food waste hydrolysate. In order to get food waste hydrolysate, vaccum filtration should be carried in 3 sequences. Filter paper (whatman number 1,Cat no.1001-110) is used for the first stage of filtration. Later the filtration of second stage is done by nylon membrane (pore size 0.45μm) and the last stage of filtration is carried out by sterilized filter capsule(0.2μm pore size). This filtrate or hydrolysate is used for fermentation (Kaur *et al.*, 2019).

11.5.1.3 Fermentation

Kaur *et al.* (2019) fermented the hydrolysate in the 2 L benchtop bioreactor which consists of a 1.15 L working capacity for fermentation process and the culture used is *Starmerella bombicola.* An inoculation loop is used for streaking a potato dextrose agar (PDA) plate which consists of potato extract (4g/l), agar (15g/l),dextrose (20g/l) and kept in incubation at 30°C for 2 days to obtain single colonies as preculture in fermentation. Single unit colonies in plates of PDA are conveyed into 5 ml food waste hydrolysate to incubate in culture tubes with round bottom. Incubation is done for 24 h at 30°C and rotated at 150 rpm. Then obtained pre-culture are transferred to food waste hydrolysate of 50 ml in 250 ml shake flasks (9% inoculation) and incubation is maintained at 30°C for 24 hr with 150 rpm.

This second pre-culture was inoculated into the bioreactor and operated at 2% (v/v). Sulphuric acid and 2.5 M NaOH is used to maintain pH at 3.5 in bioreactor. Rotational speed of 800 rpm flat blade turbine impellar is used to achieve the process of agitation while aeration is performed at the rate of 3.48 vvm (Kaur *et al.*, 2019). The temperature which need to maintain at 30°C is obtained by circulation of cold water with the use of chiller and even anti foam agents are also added, if necessary. Samples are drawn and sophorolipids concentration is quantified by using anthracenone.

11.5.2 Quantification and Characterization of Sophorolipids

Anthrone assay is used to measure sophorolipid concentration. Anthrone belongs to polycyclic aromatic hydrocarbon which is an organic compound and commonly known as Anthrocenone. Anthrone will give the sugar concentration released after samples acid hydrolysis. 0.2 ml of cell free sample is added to 18M concentrated sulphuric acid containing 1 ml of anthrone reagent in it. This mixture is heated at 92^{0}C for 30 min and then cooled in water for 5 min. Here absorbance is measured as 630 nm and the concentration of sophorolipids is estimated (Kaur *et al.*, 2019).

11.6 Conclusion

Reducing food loss and utilizing of cooked food waste is the responsibility of every citizen because it takes several months to bring the food from farm to plate but it takes only few seconds to throw it away. Every person has to think twice while throwing leftovers away. Following the steps provided in this chapter can indirectly feed the hungry people improving their nutritional status and reducing the score of global hunger index. The Bokashi method for preparation of organic manure at domestic level or at institutional level is a very good option to use leftover. Similarly sophorolipids development using the waste also utilizes the leftover to some extent. Though these methods are not using all the nutrients of the leftover food , that can only be done by making it possible to consume the leftover by the needy people.The supply chain of the leftover is to be created so that the food can be kept safe till consumption.

11.7 References

Abu-Khader MM. 2006. Recent progress in CO_2 capture/sequestration: a review. Energy Sources, 28, 1261-1279.

Ambong S, Nawawi WNW, Ramli N, Daud NM. 2016. Producing Fertilizer from Food Waste Recycling using Berkeley and Bokashi Method. Int. Sci. Res. Journal. 72. 75-83.

Bokashi - Composting Cooked Food Waste for Your Garden, https://www.growveg.com/guides/Bokashi-composting-cooked-food-waste-for-your-garden/2010.

Bulluck Iii LR, Brosius M, Evanylo GK, Ristaino JB. 2002. Organic and synthetic fertility amendments influence soil microbial, physical and chemical properties on organic and conventional farms. Applied Soil Ecology, 19(2), 147-60.

Coated Steel Market by Application & Resin Type - Global Forecast 2020. Markets and Markets Last Updated on March-2019. https://www.google.com/search?sxsrf=ALeKk01R7_97wDBmPPOcwvJpweBr7TBT7A:16082 76702566&q =Coated+ Steel+ Market+by+Application+%26+Resin+Type+-+Global+Forecast+2020+Marketsand Markets+Last+Updated+on+March-019.+. &sa=X&ved = 2ahUKEwjR5LnMgdftAh UBfisKHYcRBcwQgw N6BAgHEAE&biw=13 66&bih=568. FAO, Food Loss and

Food Waste 2018. http://www.fao.org/food-loss-and-food- waste/en/%0Ahttp://www.fao.org/policy-support/policy-themes/food-loss-food-waste/en/, 2018. Gorin PA, Spencer JF, Tulloch AP. 1961. Hydroxy fatty acid glycosides of sophorose from Torulopsis magnoliae. Canadian Journal of Chemistry, 39(4), 846-855.

Graham-Rowe E, Jessop DC, Sparks P. 2014. Identifying motivations and barriers to minimising household food waste. Resources, conservation and recycling, 84, 15-23.

Kasavan S, Mohamed AF, Halim SA. 2017. Sustainable food waste management in hotels: case study langkawi unesco global geopark. Plan. Malaysia, 15(4), 57-68.

Kaur G, Wang H, To MH, Roelants SL, Soetaert W, Lin CS. 2019. Efficient sophorolipids production using food waste. Journal of Cleaner Production, 232, 1-11.

Koppel K, Higa F, Godwin S, Gutierrez N, Shalimov R, Cardinal P, Donfrancesco B, Sosa M, Carbonell-Barrachina A, Timberg L, Chambers E. Foods, 5(3), 66, 2016.

Kulshrestha A, Richa. 2017. Shelf life extension of cooked food : a modified atmospheric approach to reduce municipal solid waste. International Journal of Current Research, 9(12), 622.

Martin-Rios C, Demen-Meier C, Gössling S, Cornuz C. 2018. Food waste management innovations in the foodservice industry. Waste management, 79, 196-206.

Mohan V, Deepak B, Mona S. 2017. Reduction and Management of Waste in Hotel Industries. Int. J. Eng. Res. Appl., 7, 34-37.

PHD Research Bureau. 2020. Global Hunger Index: India's ranking slips to 103 among 119 countries. https://www.phdcci.in/wp-content/uploads/2018/10/Global-Hunger Index-Indias- ranking-slips- to-103-among-119-countries.pdf.

Shin K, van Diepen G, Blok W, van Bruggen AH. 2017. Variability of Effective Micro-organisms (EM) in Bokashi and soil and effects on soil-borne plant pathogens. Crop Protection, 99.

US EPA, 2020. Reducing wasted food at home reduce, reuse, recycle. https://www.epa. gov/recycle/reducing-wasted-food-home.

What is Bokashi bran?, Bokashi Living. https://Bokashiliving.com/what-is-Bokashi-bran/, 2016.

Xiaohou S, Min T, Ping J, Weiling C. 2008. Effect of EM Bokashi application on control of secondary soil salinization. Water Science and Engineering. Water Science and Engineering, 1(4), 99-106.

12

Green Packaging and Utilization of Fruit Fibres: A Review

Srishti Khurana, Yashmita Grover, Aparna Agarwal*
Abhishek Dutt Tripathi* and *Anjana Kumari

ABSTRACT

Green packaging is a technique used to develop sustainable and eco-friendly packaging materials. In this review paper we have examined the data currently available on the use of plant by-products and plant extract in the development of green packaging. The use of plant fibres is especially beneficial because they often contain a high concentration of bioactive compounds. These compounds lend a preservative effect to the packaging material. Further we have also discussed the degradability of such packaging and the additives which need to be added in order to enhance their utility. Lastly, we have examined the possible applications of green packaging in the food industry.

Keywords: Green Packaging, Plant by-products, Fruit fibres, Biodegradability, Bioactive compounds, Nanocomposites, Renewable resources, Biopolymers

12.1 Introduction

Packaging is the art and science of enclosing or containing products with the aim of protecting them from physical, chemical and biological hazards during storage, distribution and sale. All foods need to be packaged in some form or the other to ensure their safe and wholesome delivery to the consumers. Thus, packaging is used for:

- Protection of food products
- Safe distribution of foods
- Storage of foods
- Informing and communicating product related information with consumers
- Presenting the products to consumers (Ojha *et al.*, 2015).

*Corresponding Author

Glass, plastics, metals, laminates and paper are the key materials used for packaging of food items. A variety of plastics, including high density polyethylene, polyolefin, polyvinyl chloride, polystyrene and ethylene vinyl alcohol, are used in different forms for packaging of foods. Because plastics care easily mouldable, heat sealable, are available in a variety of shapes, sizes and colours, provide excellent barrier properties and can be easily formed into films, they are ideal for food packaging applications (Marsh and Bugusu, 2007). These advantages of plastics over other packaging materials has led to a spurt in plastic production. However, the disposal and waste management of plastics is a major issue, because plastic poses a serious burden on environmental elements (Nagy and Kuti, 2016).

Because of the potential environmental threats, there is now a shift from plastic and other environmentally harmful packaging materials to eco-friendly ones.

One such environment friendly alternative is the use of fruit waste and their fibres for the manufacture of films, coating and pouches. Fruits and vegetables are consumed in large quantities as part of daily diet and the non-edible parts including peels are thrown away. Jackfruit, Pineapple and Mango generate 60%, 54% and 35-60% waste respectively (Shindeetal., 2019). These fruit and vegetable wastes are a rich source of fibre, carboxy methyl cellulose, starch and other substances, and therefore have a great potential for films, coatings and package formation. Natural fibres are biodegradable and non-carcinogenic and at the same time are cost effective. This usage of agricultural and food waste for the manufacture of value-added products has the potential to reduce the pressure on environment. The demand for bio-packaging is now picking up pace worldwide. Bio packaging, also called green packaging, has the objective of converting agricultural and agro-food residues into naturally biodegradable packaging material (Guillard *et al.*, 2018).

A variety of plant fibrescan be used for the manufacture of green packaging, including pineapple, banana, jackfruit, mango, jute and flex. The pineapple leaf fibre (PALF) is a major waste materialof the agricultural processing and comprises of hemicellulose (70-82%), lignin (5-12%) and ash (1.1%) (Asim et al., 2015). PALF shows excellent mechanical properties and can also be used for producing reinforced polymer composites. Cellulose, one of the most abundant natural polymers on earth, has also been receiving considerable attention due to its eco-friendly advantages and other attractive characteristics like nontoxicity, biodegradability, easy modification and excellent thermal and mechanical properties (Dai *et al.*, 2018). (Figure- 1 Waste Generation) ("05/00121 Replacement strategy for ASDEX upgrade's new control and data acquisition", 2005)

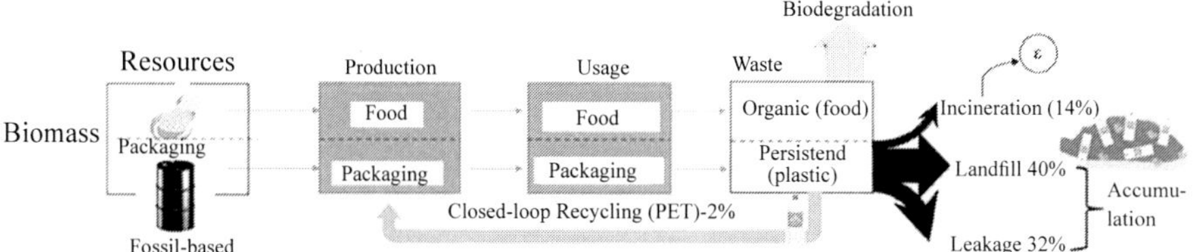

Fig. 1: Waste Generation

Nowadays, bio composite reinforced polymers are being used to replace the traditional materials owing to their light weight and high strength. These composites also exhibit high tensile and flexural strength, high compactness and high creep resistance. Natural fibres reinforced into bio-plastics, which can be considered as green packaging, can easily be degraded naturally by bacteria and enzymes. Various methods are employed to extract the natural fibres, cellulose and other materials from different food and agro-waste materials for the purpose of manufacture of bio-packaging.

12.2 Bio-based/ Green Packaging

Green packaging is a technique used to develop sustainable and eco-friendly packaging materials. These can be reused and recycled easily and utilizes the minimum amount of energy and resources as possible. This is made possible due to its Green design, which is the inclusion of environmental well- being principles in designing a product (Zhou, 2014). Bio-packaging is a term used to refer to the protective material used to cover, coat or encase the product and composed of biopolymers instead of being polymeric. Biopolymers include four general categories- 1. Proteins, 2. Lipids, 3. Polysaccharides and, 4. Polyesters (Gontard and Guilbert, 1994). These biopolymers are mostly manufactured using inexhaustible resources (Comstock *et al*., 2004; Weber *et al*., 2002). Biobased materials are mostly all biodegradable but biodegradable materials can also be from sources other than biobased (Weber *et al*., 2002). (Table 1and Figure-2 Types of Biopolymers on the basis of sources (Qasim *et al*., 2020))

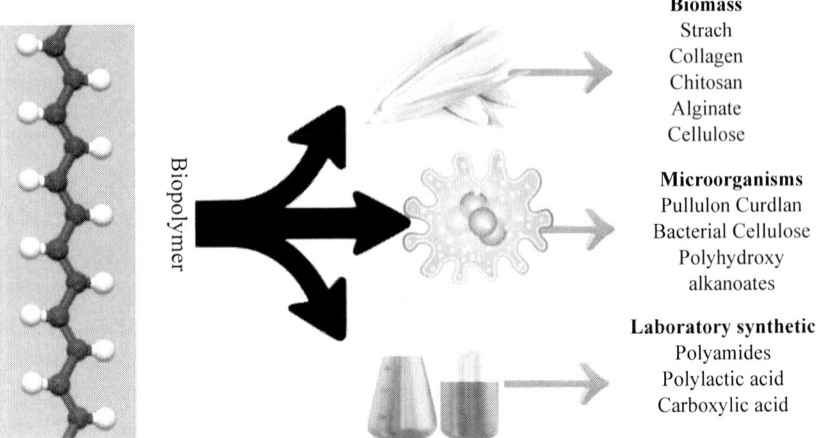

Fig. 2: Types of Biopolymers on the basis of sources (Qasim *et al.*, 2020)

Table 1: Types of Biopolymers based on source and their examples

S.No.	Type of Biopolymer	Example	Reference
1.	Proteins	1. Animal origin- Casein, Whey, ollagen, Gelatine 2. Plant Origin- Zein, Soy, Gluten	C.N. Cutter, 2006, Fig-1
2.	Polysaccharides	1. Starch- Potato, Corn, Wheat, Rice, Other derivates 2. Gums- Alginates, Pectin, Carrageenan, Guar, Locust bean etc. 3. Cellulose- Cotton, Wood, Other derivatives. 4. Chitosan/Chitin	Comstock et al., 2004; Cutter & Sumner, 2002, C.N. Cutter, 2006, Fig-1
3.	Lipids	Oils, Fats & Waxes	Comstock et al., 2004; Cutter & Sumner, 2002
4.	Polymers	1. Bio-derived polymers- Polyactate (PLA), Polyester 2. Synthesized from organisms- Pullulan, Curlan, Cellulose, Xanthan,	Comstock et al., 2004; Kandemir et al., 2005

Biodegradable plastics can be synthesized using Bioreactors with the use of bacteria, this is called — White Technology (Mooney, 2009). (Figure-3, Types of Bioplastics on the basis of their degradability (Mooney, 2009) and (Cooper, 2013))

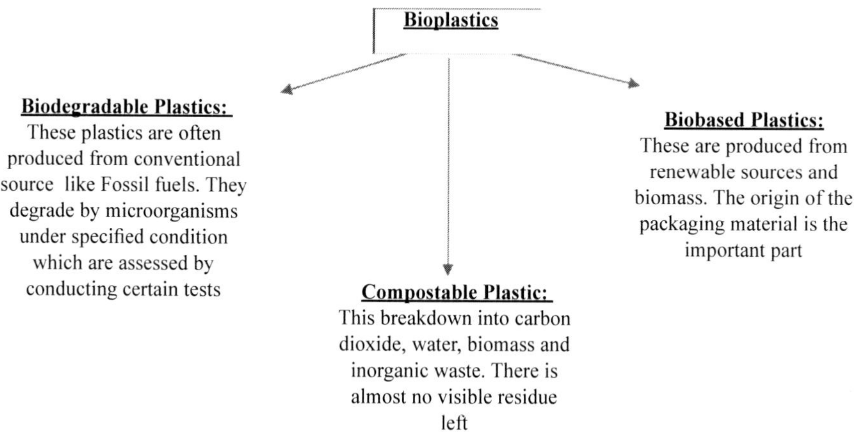

Fig. 3: Types of Bioplastics on the basis of their degradability (Mooney, 2009) and (Cooper, 2013)

12.3 Packaging Materials from Plant By-Products

A huge amount of waste is produced during the industrial processing of fruits and vegetables. It includes peels, seeds, fibres etc. This waste is a rich source of bioactive compounds like phenols, vitamins, minerals etc. (Dilucia *et al.*, 2020). This waste can be used as raw materials to synthesize polymers such as PLAs (Demichelis *et al.*, 2017). Carboxymethyl cellulose (CMC) was produced from Pineapple peels using the process of etherification. CMC is a biopolymer which can be used to manufacture bioplastics that can be used to package dehydrated food products (Chumee and Khemmakama, 2014)

Foods obtained from plants contain numerous phytochemicals and bioactive compounds. These compounds attribute antifungal, antibacterial, antiviral, anti-inflammatory, antioxidant etc. (Wang *et al.*, 2012). Addition of plant extracts to polymers can be done to improve their barrier properties and incorporate antimicrobial properties to them (Mir *et al.*, 2018). Bioactive compounds are first extracted from the plant material and then combined with film producing agents (Dilucia *et al.*, 2020). The composite film formed upon infusing banana peel extract with chitosan as an antioxidant improved the shelf life of apples post-harvest (Zhang *et al.*, 2019). The addition of lysozyme to Jackfruit-based film protected against *M. lysodeikticus.* (Bonomo *et al.*, 2017).

Another way to utilize the extracts obtained is to coat packaging materials to improve their barrier properties. One example of this is the use of citrus peels and leaves to coat kraft paper to improve its moisture and gas barrier (Kasaai and Moosavi, 2017).

12.4 Plant By-Products in Green Packaging

Various plant parts including leaves, stems, flowers, roots and seeds have long been identified to carry some vital biological compounds, exhibiting anti-bacterial, antioxidant, anti-carcinogenic and anti- fungal properties (Shima *et al.*, 2020). The incorporation of these bioactive compounds to the packaging material, or the use of such plant parts themselves to manufacture packaging material can prove to be beneficial for both the environment and food processing industries. A plethora of waste products are obtained during food processing. Some examples - pomace, seeds, pulp, husk and peels. Though most of the time these are discarded as waste, research shows that they have extensive variety of biologically active compounds and many polyphenols as well. Sometimes, the waste has more concentration of such biologically active componentsas compared to the vital parts themselves (Ayala-Zavala *et al.*, 2013). Thus, utilization of such waste products for potential packaging applications is promising from food safety and environmental point of view. The processing of fruits produces various quantities of pomace, which accounts for almosta quarter of the processed fruits (Schieber *et al.*, 2003). The pomaceincludes skin and seeds. It is rich in flavonoids, non-flavonoids, phenolic compounds and fibre(Sudha *et al.*, 2007).

12.4.1 Pineapple Peel and Fibres

Pineapple is a member of the diverse *Bromeliaceae*family. It is consumed fresh, or in the form of syrup, jam and squashes. However, the pineapple peel makes for 35% of the total pineapple. It is considered a waste during the processing and its accumulation in the environment can lead to serious environmental problems (Hu *et al.*, 2010). Hence, the pineapple peel, which is discarded as waste during processing, can potentially act as a good biomass for industrial purposes. Pineapple peels are principally rich in cellulose, hemi-cellulose, lignin and pectin (Bardiya *et al.*, 1996). From waste processing and environment protection point of view, the modification, extraction and utilization of pineapple peel cellulose are of important significance.

Cellulose is a natural polymer made of glucose units. It an abundant natural polymer and has received considerable research interest due to its environmentally friendly advantages and other features such as nontoxicity, thermal and mechanical stability, biocompatibility and easy modification. Cellulose can be utilised in the packaging sector, but infusibility and insolubility make its use limited. In order to make cellulose more processable, it is converted into its more useful derivatives including hydroxyl-ethyl cellulose, methyl-cellulose, cellulose butyrate, cellulose acetate-butyrate and cellulose acetate (Behjat, 2014). Chumeeand Khemmakama (2014) extracted cellulose from pineapple peel following a two-step process, and using the process of etherification, converted the obtained cellulose to carboxymethyl cellulose (CMC). The CMC thus produced was used as a composition of bio plastics. (Fig. 4 Pineapple leaf Fibres)

Fig. 4: Pineapple Leaf Fibres

CMC is one of the most important cellulose ether available commercially. It possesses good water solubility and is immiscible in oils and organic solvents. These properties make it a multifunction agent in the food industry and is thus used as a stabilizer, thickener and binder. Most importantly, because it forms a continuous matrix, it is used to make edible films and bio coatings for various foods. Various substances can be added to CMC to optimize its packaging properties. Sayanjali *et al.* (2011) observed the antimicrobial characteristics of CMC films after adding Potassium sorbate to it. The enhancement of tensile strength, Young's modulus &mechanical properties by the addition of essential oils was studied by Biswas *et al.* (2018). CMC films can also be purposed as edible coatings, water soluble dissolvable bags and pouches for various food items.

The Pineapple leaves are an excellent source of cellulosic fibre. After the harvest of pineapples, pineapple leaves are left with an appreciable amount of cellulosic fibre (Sibaly and Jeetah, 2017). These leaf-fibrescan be pulped to make environmental-friendlypapers and coatings. However, a drawback of cellulose fibre derived packaging material is poor resistance to water. This problem can be overcome by the application of suitable bio-coatings. Zhang *et al.*, (2017) reported that paper cups could be made waterproof upon the application of sodium alginate and gellan gum coating. Beeswax, a widely used food grade additive, is also used as a coating on packaging materials to improve water resistance. Iewkittayakornetal. (2020) obtained fibrous pineapple leaf pulp from the leaves and processed the pulp into paper. This paper was later overlayed with 30% beeswax + 4% glycerol +1% chitosan and compressed into biodegradable plates. The pulp obtained from the pineapple leaves is a potential raw material to producebio packaging.

12.4.2 Jackfruit Seed, Rind and Straw

Jackfruit (*Artocarpus heterophyllus)* is an exotic tropical fruit having a subtle sweet flavour and is consumed in many parts of the world. It is eaten raw (when ripe) and is also used to prepare sweet and savoury dishes, including desserts and curries. The most commonly consumed part of jackfruit is the flesh, also called fruit pods, which is edible in both ripe and unripe form. The inedible part, that is the rind and the straw of jackfruit are most often discarded as agricultural waste. Jackfruit waste includes skin, seeds, straw and core and accounts for approximately 60-70% of the total weight (Subburamu *et al*., 1992).

Jackfruit rind constitutes for more than half of the jackfruit, which is a major by-product (Cheok *et al*., 2018). Ooi *et al*., (2011) reported that the flour obtained after drying the jackfruit by-productscould be used in manufacturing of food packaging. Thesc bio substanccs can bc combined with suitable synthetic polymers to obtain biodegradable packaging films. Using a similar approach, Sarebanha *et al*., (2018) observed that flour obtained from jackfruit by-products is a cheap alternative which has the potential to be utilized as a packaging material. This was followed after preparing composite films from Polyvinyl alcohol and jackfruit waste flour and studying various characteristics such as film appearance and thickness, solubility, water vapor transmission, biodegradability, tensile strength and seal strength and oxygen permeability. Production of such composite films will not only allow the utilization of agricultural waste, but also will reduce the over utilization and dependency on synthetic polymers.

Another natural material which can be combined with synthetic polymers to obtain biodegradable packaging material is starch, which is found in abundance in jackfruit seeds. Though the seeds are considered edible and are consumed in some places, they are discarded as waste in most countries. These seeds are an excellent source of protein and starch. Many experiments have been performed to isolate starch from jackfruit seeds and utilize it film making. Sarifuddin *et al*., (2018) prepared PVA (poly vinyl alcohol) blend containing jackfruit seed starch via solution casting method and observed that the starch dispersed in the PVA blend effectively and had a good compatibility. It was also reported that the prepared film had an optimum tensile strength of 10.45MPa, 2.28% elongation at break and showed improvement in degradation behaviour under normal conditions in comparison to pure PVA. Retnowati *et al.*, (2015) developed biodegradable films from jackfruit seed and durian seed flours, and reported that jackfruit seed flour, owing to its higher amylose content, contributed to the strength and fracture strain of the film.

The straw of jackfruit comprises of 40-50% of jackfruit waste, which is composed of the waste jackfruit flesh without skin and seeds (Neswati *et al.*, 2015). Jackfruit straws are normally discarded as waste and their accumulation in the environment can lead to serious issues. There have been many reports suggesting the potential use of jackfruit straw for the manufacture of biodegradable films.

Jackfruit straw is a rich source of carbohydrates, like glucose, fructose, starch, sucrose, pectin and cellulose, making it a promising renewable source for biopolymers. Nazri *et al.* (2019) reported the effects on the properties of JSP (jackfruit straw powder) film upon the addition of tapioca starch and reportedthat is responsible for the increased tensile strength and modulus of JSP/starch film. (Neswati *et al.*, 2015) observed the changes when red ginger extract on the thickness, water vapor transmission rate (WVTR) and water activity of edible jackfruit straw film. Hence, jackfruit waste is a renewable material which has the potential to be utilized for the preparation of edible bio-based packaging material.

12.4.3 Mango Peel and Kernel

Mango, also called the –king of fruits is a stone fruit belonging to the genus *Mangifera.* In mango fruit, the mesocarp is fleshy and eaten, while the epicarp (forming the skin) and the endocarp (hard and stony kernel) are inedible and discarded as waste. Annually, approximately 10 lakh ton of its seeds are discarded, knowing that it makes up 20-60% of the whole fruit. The peel, constituting about 15-20% of the fruit is also discarded. This mango waste generated in large quantities is rich in various bio active compounds and hence, has the potential to be utilised in the manufacture of bio films. Thus, the shelf life of the enclosedfood can be prolonged. Working on similar lines, mango peelextract from four mango varieties (best antioxidant and antibacterial activities reported by Langra variety peel) was prepared (Kanatt *et al.*, 2017). A film was prepared using PVA-cyclodextrin-gelatineinfused with Langra Mango Peel Extract. Chicken minced was packed in the prepared films. An improvement in the mechanical and UV-barring properties of the film was observed. The chicken remained safe for consumption for 10 more days than the control after packaging in the film and storing at chilled temperatures. (Fig.-5 Mango Kernel)

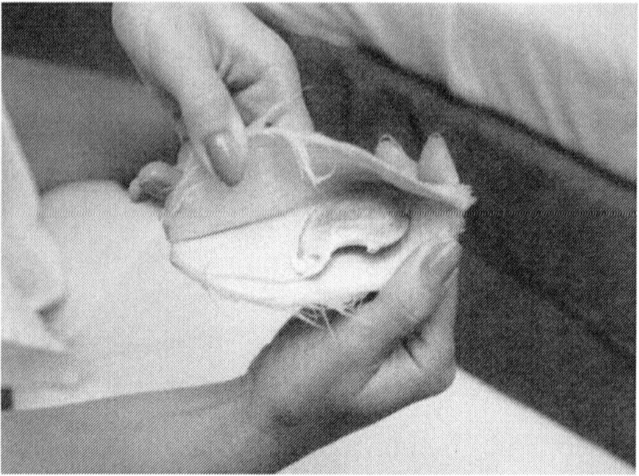

Fig. 5: Mango Kernel

Melo *et al.*, (2019) with colleagues has presented an excellent way to utilize mango kernel. The researchers have extracted 3 components from the kernel namelykernel fat, kernel phenolic extract and kernel starch. They were used to produce bioactive films. It was reported the films had UV absorbing, antioxidantand good barrier properties. It lacked strength and transparency. Rambabu *et al.*, (2018) studied the addition of Mango Leaf Extract into chitosan and used the prepared polymer for cashew nut storage. It was observed that the cashews stored in 3% and 5% films showed significant oxidation resistance.

12.4.4 Banana Peel and Fibres

Banana is an elongated edible fruit belonging to the *Musaceae* family. Annually, 102 million tons of banana is produced worldwide(Vu *et al.*, 2018). Banana peels which are a waste generated by the industry,which are packed with bioactive ingredients, including polyphenols, carotenoidsand isenriched with pro-vitamin A. They also possess antioxidants, which protect the body from free- radical damage. Faria *et al.* (2018) developed edible coating using ripe —Prata banana peels and cornstarch. The films, prepared using casting method, were biodegradable, suggesting that ripe —Prata banana peel flour is a promising ingredient which can be utilized for the manufacture of bio-films. Zhangetal., (2019) reported an interesting way to utilize banana peels. The researchers developed modified active packaging by infusing chitosan film and coating withripe banana peel extract. This was in the form of coatings and films. It was observed that the prepared chitosan-banana peel extract composite film had exceptional antioxidant properties in various food stimulants. Additionally, the developed coating led to lower respiration and weight loss of the coated apples.

12.4.5 Pomegranate Peel, Seed and Pomace

Pomegranate is a red coloured fruit which is native to Middle Eastern cuisine, and its popularity has risen worldwide. It known to have a plethora of therapeutic benefits and great taste (Pagliarulo *et al.*, 2016). The arils of this fruit are often used in industry to release juice but they can also be eaten raw like any other fruit. The industrial processing of pomegranate creates a generous amount of scrap, mainly pomegranate peels and pomace. The peel is almost half of the gross weight of the fruit(Ahmad *et al.*, 2015). The peel possesses higher biological activity among the other fruit components. The pomegranate by-products, mainly peel, seed and pomace generated by the juice and jam industries were re-used for conducting studies (Dilucia *et al.*, 2020). Recent studies indicate the use of pomegranate peel powder for the manufacture of biodegradable films. (Mushtaq *et al.*, 2018) attempted to develop of a bioactive film from zein which possessed superior mechanical properties. Pomegranate peel extract was added to the prepared Zein film mixture at various concentrations. Enhanced functional characteristics were observed when PPE was added to the zein film. The film prepared was appliedto Kalari cheese as packaging. It was observed that the film resulted in lower rates of microbial and oxidative damage to the cheese. The sensory acceptability of the Kalari cheese remained almost unchanged. In another study, Hanani *et al.*, (2018) developed active packaging films based on fish gelatinein which pomegranate peel powder (PPP) was added. It was observed that PPP was responsible for providing an appreciable antimicrobial and antioxidant effect. Its incorporation also led to a stronger film. But on the other hand, water vapor transmission rate also increased. Further studies and extended research work can result in better performing films.

12.4.6 Coconut Fibres

Coconut is a tropical fruit which is used for its water, milk, oil and the delicious meat. As the coconut ripens over time, its outer surface develops dark brown coloured husk. From this husk is extracted the coconut fibre, also called coir. These fibres are used in the manufacture of various products, such as doormats, brushes and floor mats. The green and tender coconut is also a rich source of fibre and cellulose. Cerqueira *et al.*, (2017) prepared cassava and potato starch based biodegradable films having coconut fibre derived cellulose nanocrystals incorporated in them. The nanobiocomposite biodegradable films developed showed desirable barrier and mechanical properties. The addition of cellulose extracted from the coconut fibres conferred mechanical strength to the film. (Fig.-6 Coconut Fibres)

Fig. 6: Coconut Fibres

12.5 Bioactive Substances in Green Packaging

The integration of plant extract from different sources into biodegradable films and packaging materials, changes their physicochemical and mechanical behaviours. The often improve the barrier properties of the packaging material (Bifani *et al.*, 2007; Mayachiew & Devahastin, 2010; Norajit *et al.*, 2010). The source for these extracts can be leaves, seeds, peels, fruits etc. The major factors that determine the extent of changes in the properties of the packaging material are the concentration and kind of extract used. Introduction of plant extract to the polymer can lead to changes in its thickness, water vapour permeability, colour, solubility, strength and its microbial and oxidative barrier properties. (Mir *et al.*, 2018)

The use of mango kernel extract resulted in a thicker and stronger edible film with a lower water vapour permeability and (Adilah *et al.*, 2018b). Another study where cinnamon, clove and star anise extract were applied for a gelatine-based film, the film had a lower water permeability and higher tensile strength (Hoque *et al.*, 2011). When rosemary concentrate was use with tuna skin gelatine its deformation value substantially dropped (Gomez-Estaca *et al.*, 2009)

Bioactive agents are used for their antimicrobial and antioxidant properties in green packaging. The phenolic compounds present in plant extract can act on the bacterial cell membrane and prevent microbial reproduction from taking place. These can also inhibit the activity of enzymes, linking themselves to proteins or inserting itself into the cell wall/DNA (Sivarooban *et al.*, 2008).

A combination of extracts can be used to increase potency against pathogen like the use of green tea extract, grapeseed extract and nisin to control the growth of *L.* monocytogenes (Theivendran *et al.*, 2006). In an interesting study conducted by (Aldana *et al.*,2015)essential oil extract from Mexican lemon when used as a protective coating on the fruit itself inhibited growth of *Escherichia coli*, *Salmonella thyphimurium*, *Bacillus cereus*, *Staphylococcus aureus* and *Listeria monocytogenes*.

Another factor affected by the addition of plant extract and thus bioactive substances is the change in colour. This change is sometime desirable and other times not. It does affect the sensory perception of the consumer. One good example of desirable change in colour would be the addition of cranberry pomace extract which gave a bright red colour to Soy Protein Isolate (SPI) film. (Park & Zhao, 2006)

The use of antimicrobial agents can turn detrimental if used incorrectly. When essential oils of oregano, clove and garlic are added to fish proteins they have an unfavourable impact on the tensile strength, elongation and transparency (Texeira *et al.*, 2014).

12.6 Additives used in Green Packaging

Biopolymers can be made more suitable as a packaging material by adding certain additives like plasticizers, lipids etc. which improve their barrier and mechanical properties (Castro-Rosas *et al.*, 2016). Starch is an important biopolymer used to produce edible and green packaging materials. It has many beneficial properties such as thermoplasticity and biodegradability. It is also cheap and accessible. The tensile strength and barrier properties of starch aren't sufficient by itself. Addition of plasticizers is thus done to improve its utility (González *et al.*, 2016).

Addition of Sorbitol to a Jackfruit based starch film, resulted in a mechanically and thermally stable product (Zahiruddin *et al.*, 2018).Similarly, when three sweeteners Xylitol, Sorbitol and Fructose were used as plasticisers, the water permeability rate decreased for edible film (Tong *et al.*,2013). Plasticizers also reduce brittleness and gives flexibility to films (Perez *et al.*, 2016).

Another additive being used for biopolymers are nanoparticles. Nanocomposites thus formed have better mechanical and barrier properties. The nanoparticles being added are mostly nanofillers. Castro-Rosas *et al.*, 2016).

Lipids can be used to enhance the moisture barrier properties of polysaccharides and protein-based films. They increase hydrophobicity of the film and thus lower the permeability rate of water vapours (Kowalczyk & Baraniak, 2014; Rocca-Smith *et al.*, 2016). (Figure- 7 Manufacturing of Nanocomposites, Figure- 8 Nanoparticles used in Nanocomposites).

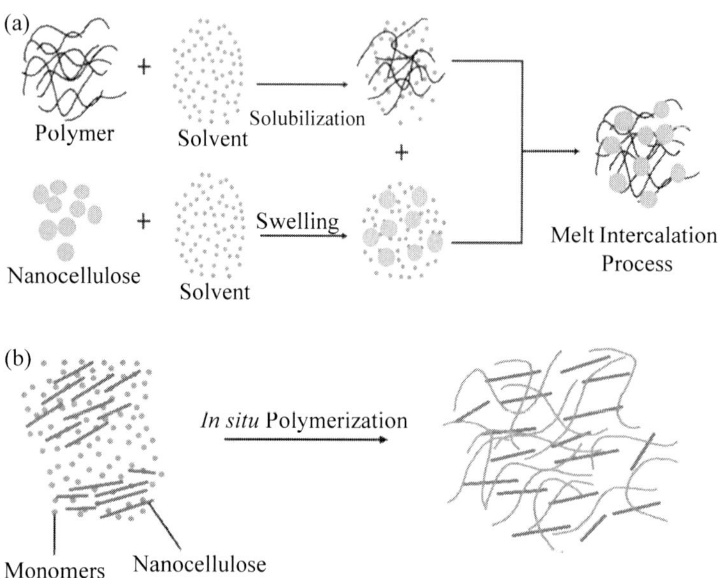

Fig. 7: Manufacturing of Nanocomposites (Qasim *et al.*, 2020)

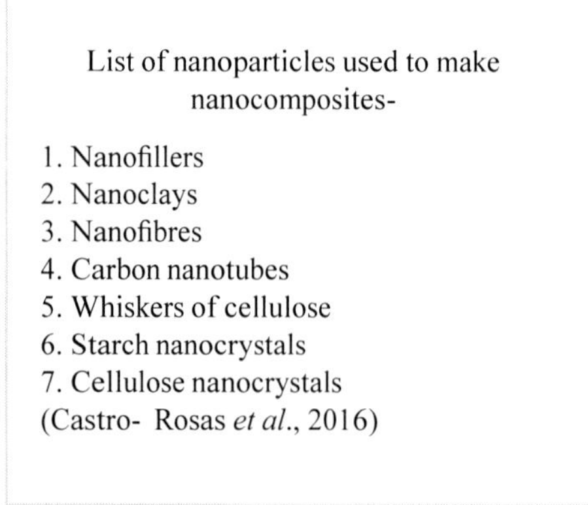

List of nanoparticles used to make nanocomposites-

1. Nanofillers
2. Nanoclays
3. Nanofibres
4. Carbon nanotubes
5. Whiskers of cellulose
6. Starch nanocrystals
7. Cellulose nanocrystals

(Castro- Rosas *et al.*, 2016)

Fig. 8: Nanoparticles used in Nanocomposites

12.7 Properties of Green Packaging

In today's modern world, great importance is being given to products prepared from renewable sources, owing to their less harmful impact on nature. Plastic as a packaging material has gained immense applications all over the world, with different forms and types of plastics being used for the packaging of different foods. However, with the increasing accumulation of plastics in the environment, increasing levels of all forms of pollution, overflowing landfills and dwindling oil reserves, efforts are being made to develop and use biodegradable green packaging materials in place of plastics. Green packaging includes bio plastics, biopolymers, Biodegradable microbial polymers and bioactive films and coatings. As discussed in the above sections, food materials discarded as waste by the processing industries can be utilized in the manufacture of such green packaging materials. The properties of the green packaging are thus influenced by the materials incorporated in it.

12.7.1 Barrier Properties

The most widely used bio materials such as paper and cellulose films have poor barrier properties, especially in terms of moisture resistance. These substances thus need to be mixed with suitable synthetic substances to improve their moisture resistance properties. Cellulose originally is sparingly water soluble, thus unsuited for forming films and coatings. In order to make cellulose water soluble, coating and blending along with surface modification methods are applied. Nanocellulose fibres obtained by the acid hydrolysis or mechanical grinding of cellulose are suitable for making films and composites. Fibres ~15-20nm thick can be utilized for film and composite making and the incorporation of these nanocellulose fibres into a cellulose polymer mix lead to better moisture barrier properties maintaining its biodegradability (Pandey *et al*., 2005).

Starch, the most commonly used biodegradable polymer in the development of films, has poor oxygen transmission rates. Brittleness and hydrophilicity are two factors that sound be kept in mind when talking about starch (Zhong *et al*., 2019). These disadvantages limit the utilization of starch for packaging of foods (Weberetal., 2002). To increase its flexibility and improve its plasticizing ability, various plasticizers like glycerol, sorbitol and glycol are added to starch with application of heat and shear, thereby converting it into Thermoplastic Starch (TPS) (Isottonetal., 2015; Abdorreza *et al*., 2011). One disadvantage of TPS is that it gets brittle on long storage. Increasing brittleness on storage can be attributed to starch retrogradation.

Proteins are widely used for film and coating making applications, and they have been proved effective in many studies. Various types of protein-based edible films are being researched, including, soy protein films,corn zein films, collagen films, wheat gluten films, gelatine films and casein films. These films offer better gas barrier properties in comparison to the polysaccharide and lipid-based films (Wittaya, 2012). Cuq *et al.*, (1998) observed that when the edible soy protein film was dry, its oxygen permeability was 50, 260, 540 and 670 times lower as compared to Low-Density Polyethylene, methylcellulose, starch and pectin respectively. The proteins obtained from different plant sources (for film making purposes) have different amino acid composition and hence the prepared films differ in oxygen permeability values (Reichert *et al.*, 2020). Whey/pea protein isolatecoatings developed by Chang *et al.*, (2019) possess a weak oxygen barrierthan multilayer films with nylon that prevents oxygen transmission. Protein films maybe altered by various treatments in order to improve barrier properties. For instance, Weng and Zheng (2015) reported greater oxygen and vapor barrier properties in films containing modified soy proteins.The modification of protein films by fatty acids can be used as a method to restrict water vapour transmission in such films. Coating of protein films with a hydrophobic material such as beeswax can also reduce the rate of water vapor transmission.

Polylactic acid, PLA, is a lactic acid-based polymer obtained from sugarcane bagasse, corn stover, and rice hulls. PLA enjoys wide acceptability as a biopackaging material due to its properties such as high transparency, easy processing, rigidity and biocompatibility (Zhong *et al.*, 2019). Brittleness and poor gas and UV barrier are disadvantages posed by PLA films. It has a moisture vapor transmission rate 3-5 times greater than that of PET (Polyethylene terephthalate) and has improved barrier properties to oxygen from PS (polystyrene) but not as well as PET (Ivanković *et al.*, 2017). In PLA films, the degree of crystallinity affects its water vapor permeability. Thus, many studies have been conducted to alter the degree of crystallization and ultimately to modify biodegradation properties and water vapor transmission of PLA films.

12.7.2 Mechanical Properties

Mechanical properties of most organic materials don't differ much from petroleum derived ones. They are defined by the molecular mass of the polymer chain and the degree of crystallization. For instance, orientation of PLA enhances the film mechanical strength and heat stability, while a different molecular weight and crystallization will result in a soft, elastic and high strength material. In PLA and derived blends, the selection and application of an appropriate plasticizer allows the modulation of biodegradability (Bugnicourt *et al.*, 2014). Using LA oligomers as plasticizers in PLA films helps enhance mechanical properties and biodegradability. PLA, is superior to PLC and PHA, since it possesses better thermal processing properties, making it suitable for processes like injection molding and blow filming (Rasal *et al.*, 2010). PLA is brittle, with elongation at break less than 10%. Ayana *et al.*, (2014) blended PLA with extremely exfoliated clay and thermoplastic starch. They reported that the obtained material had improved mechanical strength.

Protein edible films exhibit improved mechanical properties in comparison to fat and polysaccharide- based films. This can be attributed to the unique structure of proteins, which confers them high intermolecular binding potential and other functional properties. (Fig. 9 Cellulose in Green Packaging (Qasim *et al.*, 2020))

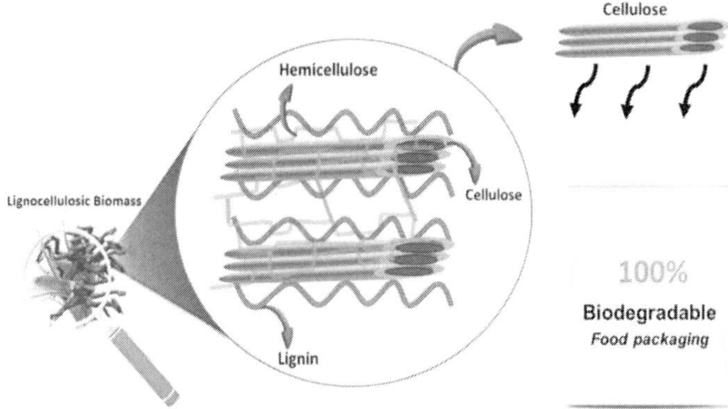

Fig. 9: Cellulose in Green Packaging (Qasim *et al.*, 2020)

12.8 Degradation/ Biodegradability

Studying the degradation of polymers is not an easy task. It is a complicated process involving the action of enzymes and other actions such as Biodeterioration, Biofragmentation and Assimilation. Here during biodeterioration due to the action of various microbes present in the environment, degradation on the surface layers takes place. During biofragmentation involves the size reduction of the polymer by chemical, photolytic, thermolytic, mechanical or biological means. During Assimilation the molecules released from the polymers after their breakdown can be used as food by microbes to feed on. (Lucas *et al.*,2008) (Figure-10 Steps involved in degradation of Polymers (Lucas *et al.*, 2008))

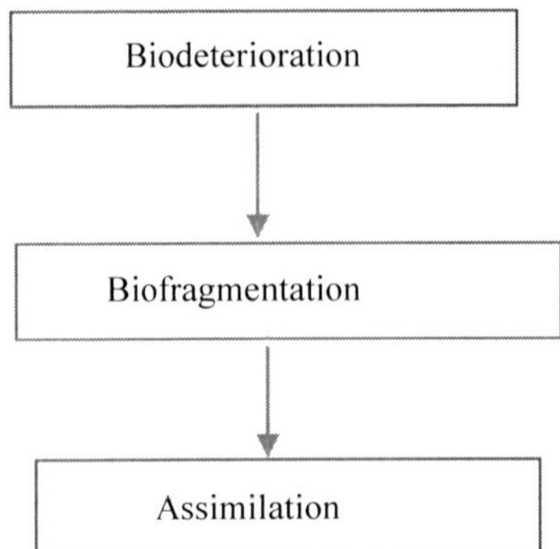

Fig. 10: Steps involved in degradation of Polymers

Many factors affect the biodegradation of a polymer. Some of them include-chemical and physical. Their environment also plays an essential role. In developing countries, a large percentage of the total waste produced ends up into landfills. This leads the emission of greenhouse gases and leaching of harmful chemicals into the soil and water. (Emadian *et al.*, 2016).

Studies conducted on decomposition of bioplastics show that decomposition is relatively very slow in the case of composting at home. The higher temperature available at an industrial level helps in hastening the process (Rudnik and Briassoulis, 2011).

The biodegradability of a material can be tested by various methods that have been validated of agencies such as ISO and ASTM. Many of these tests involve the composting under specific conditions in a laboratory setup. When the disintegration test for a whey based edible layer was conducted, it was found that only 10% of the original amount was present after 30 days (Cinelli *et al.*, 2014). To test the biodegradability of a film made out of cassava starch, (Ezeoha and Ezenwanne, 2013) used the Total Organic Carbon Test procedure. It was found that the prepared contained more organic content than polythene and lesser than paper. It can thus, get degraded better than polythene. (Table 2 Tests to check Biodegradability and Composability of Plastics)

Table 2: Tests to check Biodegradability and Composability of Plastics

S.No.	Type of Bioplastic	Test Standard	Reference
1.	Biodegradable Plastic	ASTM D5511and ASMT D5526(anaerobic); ASTM D5338 and ASTM D5209(aerobic)	Cooper, 2013
2.	Compostable Plastic	ASTM D-6400(U.S.A) and EN 13432(Europe) Carbon-14 tested by ASTM D6866	Cooper, 2013

12.9 Green Packaging in the Food Industry

Edible films incorporated with bioactive compounds can be used in the food industry on a large scale with goals, such as being consumable and possessingantioxidativeand antimicrobial compounds. These edible films can be utilised for the packaging of fresh fruits, meat,vegetables and nuts has received considerable interest. Edible films have the potential to act as carriers of bioactive substances which can prevent the spoilage of foods, keep the food fresh and may increase its shelf life. Films incorporated with antimicrobials release antimicrobial substances, which inhibit microbial spoilage by extending their lag phase or minimising their log phase (Quintavalla and Vicini, 2002). The plant derived antimicrobial substances added to the edible films are slowly released on the food surface that the film is in contact with, and thus, small concentrations of the bioactive compound are required to prolong their shelf life (Ye *et al.*, 2008). Many studies have been conducted involving incorporation of plant bio-active compounds to edible films and positive results were observed. In one such study, Ayala-Zavala etal., (2013) added leaf oil of cinnamon to edible film and observed that the film exhibited antimicrobial properties. It was reported that the film is useful inpreserving fresh-cut peaches. Similarly, Jang *et al.*, (2011) reported that rapeseed protein gelatine films incorporated with grapefruit seed extract prevented the growth of *Escherichia coli* and *Listeria monocytogenes* in the case of strawberries. Hence, the addition of small amounts of bioactive compounds exhibiting antimicrobial activity to edible films will not only protect the food from microbial spoilage, but will also substantially reduce the need for addition of antimicrobials in foods (Bahrami *et al.*, 2020). (Table-3 Applications of Green Packaging examples)

Table 3: Applications of Green Packaging examples

S.No.	Type of film	Property	Product Packaged	Reference
1.	Edible film+ cinnamon leaf oil	Antimicrobial property	Fresh cut peaches	(Ayala-Zavala et. al 2013)
2.	Rapeseed protein gelatin film+ grapefruit seed extract	Inhibited growth of *E. coli* and *L. monocytogenes*	Strawberries	(Jang *et al.* 2011)
3.	Apple based edible film+ carvacrol + cinnamaldehyde	Antimicrobial effect against- 1. *L. monocytogenes* 2. *E. coli* and *S. enterica*	For 1.Ham 2.Poultry	(Ravishankar *et al.*, 2009)

4.	Polypropylene/ ethylene vinyl alcohol copolymer films + 5% oregano essential oil+	Antimicrobial effect on pathogens like *E.coli* and *S.enterica*	Packaged Salads	(Muriel- Galet *et al.*, 2012)
5.	Chitosan coated film+ 4% green Tea extract	Inhibited *L. monocytogenes*	Ham Steak at the time of storage (refrigerated and ambient temperature)	(Vodnar and Socaciu, 2012)

12.10 Conclusion

The purpose of food packaging is not only limited to the containment of food products but it also acts an essential barrier which protects the food material from hazards and ensures its safety. Plastics have been dominating the world of food packaging for a number of decades now. This has led to a big problem of excessive waste generation and growing dependency on non-renewable sources such as fossil fuels. In order to combat this problem, the development of green packaging is ideal and is picking pace. The integration of plant extracts in food packaging can impact its antimicrobial, antioxidant and other chemical and physical properties. Plant by-products such a plant-fibres which are conventionally discarded as waste hold an opportunity to be converted into food packaging. This will not only help to manage the waste but also provide an alternative to plastic packaging. The addition of bioactive compounds in the food packaging material has the ability to prolong the shelf stability of food while lending beneficial properties to the films and coatings. With the increasing concern related to food safety amongst the consumers, green packaging holds a lot of promise. To the best of our knowledge plant by-products are a potential raw material for producing food packaging.

12.11 References

Abdorreza MN, Cheng LH, Karim AA. 2011. Effects of plasticizers on thermal properties and heat sealability of sago starch films. Food Hydrocoll., 25, 56-60.

Adilah AN, Jamilah B, Noranizan MA, Nur Hanani ZA. 2018. Utilization of mango peel extracts on the biodegradable films for active packaging. Food Packag. Shelf Life, 16, 1-7.

Ahmad M, Hani NM, Nirmal NP, Fazail FF, Mohtar NF, Romli SR. 2015. Optical and thermo-mechanical properties of composite films based on fish gelatine/rice flour fabricated by casting technique. Prog. Org. Coat, 84, 115-127.

Aldana DS, Andrade-Ochoa S, Aguilar CN, Contreras-Esquivel JC, Nevárez-Moorillón GV. 2015. Antibacterial activity of pectin-based edible films incorporated with Mexican lime essential oil. Food Control. 50, 907– 912.

Argüello-García E, Solorza-Feria J, Rendón-Villalobos JR, Rodríguez-González F, Jiménez-Pérez A, Flores-Huicochea E. 2014. Properties of edible films based on oxidized starch and zein. Int. J. Polym. Sci, 1-9.

Asim M, Abdan K, Jawaid M, Nasir M, Dashtizadeh Z, Ishak MR, Hoque ME. 2015. A review on pineapple leaves fiber and its composites. Int. J. Polym. Sci, 6.

Ayala-Zavala JF, Silva-Espinoza BA, Cruz-Valenzuela MR, Leyva JM, Ortega-Ramírez LA, Carrazco-Lugo DK, Pérez-Carlón JJ, Melgarego-Flores BG, González-Aguliar GA, Miranda MRA. 2013. Pectin-cinnamon leaf oil coatings add antioxidant and antibacterial properties to fresh- cut peach. Flavour Frag. J, 28(1), 39-45.

BA,Suin S, Khatua BB. 2014. Highly exfoliated eco-friendly thermoplastic starch (TPS)/ poly (lactic acid) (PLA)/clay nanocomposites using unmodified nanoclay. Carbohydr. Polym,110, 430-439.

Bahrami A, Delshadi R, Assadpour E, Jafari SM, Williams L. 2020. Antimicrobial-loaded nanocarriers for food packaging applications. Adv. Colloid Interface Sci, 278, 102140.

Bardiya N, Somayaji D, Khanna S. 1996. Biomethanation of banana peel and pineapple waste. Bioresour. Technol, 58(1), 73-76.

Bifani V, Ramirez C, Ihl M, Rubilar M, Garcia A, Zaritzky N. 2007. Effects of murta (Ugnimolinae Turcz) extract on gas and water vapor permeability of carboxymethylcellulose- based edible films.Lebensmittel-Wissenschaft und -Technologie- Food Science and Technology, 40(8), 1473–1481.

Biswas A, Furtado RF, Bastos MSR, Benevides SD, Oliveria MA, Boddu V, Cheng HN. Preparation and characterization of carboxymethyl cellulose films with embedded essential oils. J. Mater. Sci, 7(4), 2018.

Bonomo R, Santos T, Santos L, da Costa Ilhéu Fontan R, Rodrigues L, dos Santos Pires A, Veloso C, Gandolfi O, Bonomo P. 2017. Effect of the Incorporation of Lysozyme on the Properties of Jackfruit Starch Films. J Polym Environ, 26(2), 508-517.

Bugnicourt E. Polyhydroxyalkanoate (PHA): 2014. Review of synthesis, characteristics, processing and potential applications in packaging. Express Polym. Lett,8, 791-808.

Castro-Rosas J, Cruz-Galvez AM, Gomez-Aldapa CA, Falfan-Cortes RN, Guzman-Ortiz FA, Rodríguez-Marín ML, 2016. Biopolymer films and the effects of added lipids, nanoparticles and antimicrobials on their mechanical and barrier properties: a review. Int J Food Sci Technol, 51: 1967-1978.

Cerqueira JC, Penha JDS, Oliveira RS, Guarieiro LLN, Melo PDS, Viana JD, Machado BAS. 2017. Production of biodegradable starch nanocomposites using cellulose nanocrystals extracted from coconut fibres. Polímeros, 27(4).

Chang Y, Joo E, Song HG, Choi I, Yoon CC, Choi YJ, Han J. 2019. Development of protein-based high-oxygen barrier films using an industrial manufacturing facility. J. Food Sci,84, 303-310.

Cheok CY, Adzahan NM, Rahman RA, Abedin NHZ, Hussain N, Sulaiman R, Chong GH. 2018. Current trends of tropical fruit waste utilization. Crit. Rev. Food Sci, 58(3), 335-361.

Chumee J, Khemmakama P. 2014. Carboxymethyl cellulose from pineapple peel: Useful green bioplastic. Adv. Mater. Res,979, 366-369.

Cinelli P, Schmid M, Bugnicourt V, WildnerJ, BazzichiA, AnguillesiI, LazzeriA. 2014. Whey protein layer applied on biodegradable packaging film to improve barrier properties while maintaining biodegradability, Polym.Degrad. Stab., Volume 108, Pages 151-157, ISSN 0141-3910.

Cuq B, Gontard N, Guilbert S. 1998. Proteins as agricultural polymers for packaging production. Cereal Chem, 75(1), 1-9.

Cutter C. 206. Opportunities for bio-based packaging technologies to improve the quality and safety of fresh and further processed muscle foods. Meat Sci, 74(1), 131-142.

CutterCN, SumnerSS. 2002. Application of edible coatings on muscle foods,in: Protein-based films and coatings,A. Gennadios (Ed.), pp. 467–484,CRC Press.

Dai H, Ou S, Huang Y, Huang H. 2018. Utilization of pineapple peel for production of nanocellulose and film application.Cellulose, 25, 1743-1756.

De FariaArquelau PB, Silva VDM, Garcia MAVT, de Araújo RLB, Fante CA. 2019. Characterization of edible coatings based on ripe —Prata-banana peel flour. Food Hydrocoll, 89, 570-578.

Dilucia F, Lacivita V, Conte A, Del Nobile MA. 2020. Sustainable use of fruit and vegetable by- products to enhance food packaging performance. Foods, 9(7), 857.

Emadian S, Onay T, Demirel B. 2017. Biodegradation of bioplastics in natural environments. Waste Manage, 59, 526-536.

Ezeoha SL, Ezenwanne JN. 2013. Production of Biodegradable Plastic Packaging Film from Cassava Starch, IOSR Journal of Engineering, 3(10), 14-20.

Gomez-Estaca J, Bravo L, Gomez-Guillen MC, Aleman A, Montero P. 2009a. Antioxidant properties of tuna-skin and bovine-hide gelatin films induced by the addition of oregano and rosemary extracts. Food Chem, 112(1), 18–25.

Gomez-Estaca J, Gimenez B, Montero P, Gomez-Guillen M C. 2009b. Incorporation of antioxidant borage extract into edible films based on sole skin gelatin or a commercial fish gelatin. J. Food Eng, 92(1), 78–85.

Gomez-Estaca J, Montero P, Fernandez-Martin F, Aleman A, Gomez-Guillen MC. 2009c. Physical and chemical properties of tuna-skin and bovine-hide gelatin films with added aqueous oregano and rosemary extracts. Food Hydrocoll, 23(5), 1334–1341.

Gontard N, Guilbert S. 1994. Bio-packaging: technology and properties of edible and/or biodegradable material of agricultural origin,in: Food Packaging and Preservation, Mathlouthi M. (Ed.), pp-159-181,Springer, Boston, MA.

González P, Medina C, Famá L, Goyanes S. 2016. Biodegradable and non-retrogradable eco-films based on starch – glycerol with citric acid as crosslinking agent.Carbohydr. Polym, 138, 66–74.

Guillard V, Gaucel S, Fornaciari C, Coussy HA, Buche P, Gontard N. 2018. The next generation of sustainable food packaging to preserve our environment in a circular economy context. Front. Nutr, 5:121.

Hanani ZAN, Yee FC, Nor-Khaizura MAR. 2019. Effect of pomegranate (Punica granatum L.) peel powder on the antioxidant and antimicrobial properties of fish gelatine films as active packaging. Food Hydrocoll, 89, 253-259.

Hoque MS, Benjakul S, Prodpran T. 2011. Properties of film from cuttlefish (Sepia pharaonis) skin gelatin incorporated with cinnamon, clove and star anise extracts. Food Hydrocoll, 25(5), 1085–1097.

Hu X, Hu K, Zeng L, Zhao M, Huang H. 2010. Hydrogels prepared from pineapple peel cellulose using ionic liquid and their characterization and primary sodium salicylate release study. Carbohydr. Polym, 82(1), 62-68, 2010.

Iewkittayakorn J, Khunthongkaew P, Wongnoipla Y, Kaewtatip K, Suybangdum P, Sopajarn A. 2020. Biodegradable plates made of pineapple leaf pulp with bio coatings to improve water resistance. J. Mater. Res. Technol, 2020.

Isotton FS, Bernardo GL, Baldasso C, Rosa LM, Zeni M. 2015. The plasticizer effect on preparation and properties of esterified corn starch films. Ind. Crops Prod, 76, 717-724.

Ivanković A, Zeljko K, Talić S, Bevanda AM, Lasić M. 2017. Biodegradable packaging in the food industry. J. Food Safety Food Qual, 68(2), 23-52.

Jafarzadeh S, Jafari SM, Salehabadi A, Nafchi AM, Uthaya US, Khalil H.P.S.A. 2020. Biodegradable green packaging with antimicrobial functions based on the bioactive compounds from tropical plants and their by-products. Trends Food Sci Technol, 100, 262-277.

Jang S, Shin Y, Song KB. 2011. Effect of rapeseed protein-gelatin film containing grapefruit seed extract on _Maehyang' strawberry quality. Int. J. Food Sci. Technol,46(3), 620-625.

K R, G B, Banat F, Show PL, Cocoletzi HH. 2019. Mango leaf extract incorporated chitosan antioxidant film for active food packaging. Int J Biol Macromol.126:1234-1243.

Kanatt SR, Chawla SP. 2017. Shelf life extension of chicken packed in active film developed with mango peel extract. J. Food Saf, 38(1), e12385.

Kandemir N, Yemenicioglu A, Mecitoglu C, Elmaci ZS, Arslanoglu A, Goksungur MY, Baysal T. 2005. Production of antimicrobial films by incorporation of partially purified lysozyme into biodegradable films of crude exopolysaccharides obtained from Aureobasidium pullulans fermentation. Food Technol. Biotechnol, 43. 343–350.

Kasaai M,Moosavi A. 2017. Treatment of Kraft paper with citrus wastes for food packaging applications: Water and oxygen barrier properties improvement. Food Packag. Shelf Life, 12, 59-65.

Kowalczyk D, Baraniak B. 2014. Effect of candelilla wax on functional properties of biopolymer emulsion films-A comparative study. Food Hydrocoll, 41, 195– 209.

Lucas N, Bienaime C, Belloy C, Queneudec M, Silvestre F, Nava-Saucedo JE. 2008. Polymer biodegradation: Mechanisms and estimation techniques – A review. Chemosphere, 73(4), 429- 442.

Marsh K,Bugusu B. 2007. Food Packaging—Roles, Materials, and Environmental Issues. J. Food Sci., 72: R39-R55.

Mayachiew P, Devahastin S. 2010. Effects of drying methods and conditions on release characteristics of edible chitosan films enriched with Indian gooseberry extract. Food Chem. 118(3), 594–601.

Melo PEF, Silva APM, Marques FP, Ribeiro PRV, Souza Filho MM, Brito ES, Lima JR,Azeredo HMC. 2019. Antioxidant films from mango kernel components. Food Hydrocoll. 95, 487-495.

Mir S, Dar B, Wani A, Shah M. 2018. Effect of plant extracts on the techno-functional properties of biodegradable packaging films. Trends Food Sci & Technol., 80, 141-154.

Mooney B. 2009. The second green revolution? Production of plant-based biodegradable plastics. Biochem. J., 418(2), 219-232.

Muriel-Galet V, Cerisuelo JP, López-Carballo G, Lara M, Gavara R, HernándezMuñoz P. Development of antimicrobial films for microbiological control of packaged salad. Int. J. Food Microbiol., 157(2), 195–201. 2012.

Mushtaq M, Gani A, Gani A, Punoo HA, Masoodi FA. 2018. Use of pomegranate peel extract incorporated zein film with improved properties for prolonged shelf life of fresh Himalayan cheese (Kalari/kradi). Innov. Food Sci.Emerg., 48, 25-32.

Nagy Á,Kuti R. 2016. The environmental impact of plastic waste incineration.AARMS., 15(3) 2016:231-237.

Nazri MSM, Tawakkal ISMA, Khairuddin N, Talib RA, Othman SH. 2019. Characterization of jackfruit straw-based films: Effect of starch and plasticizer contents. PertanikaJSciTechnol, 27 (S1):1-14..

Neswati Murtius WS, Prastica A. 2015. Characteristics of jackfruit straw's edible film enriching by gingers red (Zingiber officinale, Rosc.). IJASEIT., 5(2).

Norajit K, Kim KM, Ryu GH. 2010. Comparative studies on the characterization and antioxidant properties of biodegradable alginate films containing ginseng extract. J. Food Eng., 98, 377– 384.

Ojha A, Sharma A, Sihag M, Ojha S. 2015. Food packaging- materials and sustainability- A review. Agric. Rev., 36(3) 2015:241-245.

Ooi Z, Ismail H, Abu Bakar A, Aziz N. 2011. Effects of jackfruit waste flour on the properties of poly (vinyl alcohol) film. J. VinylAddit. Technol., 17: 198-208.

Pagliarulo C, De Vito V, Picariello G, Colicchio R, Pastore G, Salvatore P, Volpe MG. 2016. Inhibitory effect of pomegranate (Punica granatum L.) polyphenol extracts on the bacterial growth and survival of clinical isolates of pathogenic Staphylococcus aureus and Escherichia coli. Food Chem., 190, 824-831.

Pandey JK, Kumar AP, Misra M, MohantyAK, Drzal LT, Singh RP. 2005. Recent advances in biodegradable nanocomposites. Journal of Nanosci. and Nanotechnology, 5, 497-526.

Park S, Zhao YY. 2006. Development and characterization of edible films from cranberry pomace extracts. J. Food Sci., 71(2), E95–E101.

Pleissner, D, Demichelis F, Mariano S, Navarro Gutiérrez IM, Schneider R, Venus J. 2017. Direct production of lactic acid based on simultaneous saccharification and fermentation of mixed restaurant food waste. J Clean Prod143:615-623.

Quintavalla S, Vicini L. 2002. Antimicrobial food packaging in meat industry. Meat Sci.62(3), 373- 380.

Rasal RM, Janorkar AV, Hirt DE. 2010. Poly (lactic acid) modifications. Prog. Polym. Sci., 35, 338- 356.

RavishankarS, Zhu L, Olsen CW, McHugh TH, Friedman M. 2009. Edible apple film wraps containing plant antimicrobials inactivate foodborne pathogens on meat and poultry products. J. Food Sci., 74(8), M440–M445.

Reichert CL, Bugnicourt E, Coltelli MB, Cinelli P, Lazzeri A, Canesi I, Braca F, Martinez BM, Alonso R, Agostinis L, Verstichel S, Six L, De Mets S, Gómez EC, Ißbrücker C, Geerinck R, Nettleton DF, Campos I, Sauter E, Pieczyk P, Schmid M. 2020.Bio-based packaging: Materials, modifications, industrial applications and sustainability. Polymers,12(7), 1558.

Retnowati DH, Ratnawati R,Purbasari A. A biodegradable film from jackfruit (Artocarpus heterophyllus) and durian (Duriozibethinus) seed flours. 2015. Scientific Study & Research: Chemistry & Chemical Engineering, Biotechnology, Food Industry,16(4), 395-404.

Rocca-Smith J, Marcuzzo E, Karbowiak T, Centa J, Giacometti M, Scapin F, Venir E, Sensidoni A, Debeaufort F. 2016. Effect of lipid incorporation on functional properties of wheat gluten based edible films. J. Cereal Sci. 69, 275– 282.

Ruban, S. Biobased Packaging. 2009. Application in Meat Industry. Vet.World, 2, 79-82.

Rudnik E, Briassoulis D. 2011. Degradation behaviour of poly (lactic acid) films and fibres in soil under Mediterranean field conditions and laboratory simulations testing. Ind. Crops Prod. 33, 648–658.

Sarebanha S, Farhan A. 2018. Ecofriendly composite films based on polyvinyl alcohol and jackfruit waste flour. JPackageTechnolRes, 2(3), 181-190.

Sarifuddin N, Shahrim NA, Rani NNSA, Zaki HHM, Azhar AZA. 2018. Preparation and characterization of jackfruit seed starch/poly (vinyl alcohol) (PVA) blend film. IOP conf. ser., Mater. sci. eng.290. 012065.

Sayanjali S, Ghanbarzadeh B,Ghiassifar S. 2011. Evaluation of antimicrobial and physical properties of edible film based on carboxymethyl cellulose containing potassium sorbate on some mycotoxigenic Aspergillus species in fresh pistachios. LWT- Food Sci. Technol.44, 1133-1138.

Schieber A, Hilt P, Streker P, Endreß HU, Rentschler C, Carle R. 2003. A new process for the combined recovery of pectin and phenolic compounds from apple pomace. Innov. Food Sci. Emerg. Technol.,4(1), 99-107.

Shinde M, Sonawane SK, Patil S. 2019. Fruit peel utilization in food packaging. Indian Food Industry Magazine, 1(5).

Sibaly S,Jeetah P. 2017. Production of paper from pineapple leaves. J. Environ. Chem. Eng., 5(6), 5978-5986.

Sivarooban T, Hettiarachchy NS, Johnson MG. 2008. Physical and antimicrobial properties of grape seed extract, nisin, and EDTA incorporated soy protein edible films. Food Res. Int., 41(8), 781– 785.

Subburamu K, Singaravelu M, Nazar A,Irulappan I. 1992. A study on the utilization of jackfruit waste. Bioresour.Technol., 40(1), 85-86.

Sudha ML, Baskaran V,Leelavathi K. 2007. Apple pomace as a source of dietary fibre and polyphenols and its effect on the rheological characteristics and cake making. Food Chem.,104(2), 686-692.

TA Cooper. 2013. Developments in bioplastics materials for packaging food, beverages and other fast-moving consumer goods, in: Trends in Packaging of Food, Beverages and Other Fast- Moving Consumer Goods (FMCG), N Farmer (Ed.), pp.108-152, Woodhead Publishing, 2013.

Tajeddin B. 2015. Cellulose-based polymers for packaging applications, in: Lignocellulosic Polymer Composites: Processing, Characterisation, and Properties,VK Thakur (Ed.), pp. 477-498, Scrivenr Publishing LLC.

Texeira B, Marques A, Pires C, Ramos C, Batista I, Saraiva JA, Nunes ML. 2014. Characterization of fish protein films incorporated with essential oils of clove, garlic and origanum: physical, antioxidant and antibacterial properties. LWT, 59, 513–539.

Theivendran S, Hettiarachchy NS, Johnson, MG. 2006. Inhibition of Listeria monocytogenes by nisin combined with grape seed extract or green tea extract in soy protein film coated on turkey frankfurters. J. Food Sci., 71(2), M39–M44.

Tong Q, Xiao Q, Lim LT. 2013. Effects of glycerol, sorbitol, xylitol and fructose plasticisers on mechanical and moisture barrier properties of pullulan–alginate–carboxymethylcellulose blend films. Int J Food Sci Technol, 48: 870-878.

Vodnar DC, Socaciu C. 2012. Green tea increases the survival yield of Bifidobacteria in simulated gastrointestinal environment and during refrigerated conditions. Chem. Cent. J. 6(1), 61.

Vu HT, Scarlett CJ,Vuong, Q.V. 2018. Phenolic compounds within banana peel and their potential uses: A review. J. Funct. Foods, 40, 238-248.

Wang S, Marcone M, Barbut S, Lim L. 2012. Fortification of dietary biopolymers-based packaging material with bioactive plant extracts. Food Res. Int., 49(1), pp.80-91.

Weber CJ, Haugaard V, Festersen R, Bertelsen G. 2002. Production and applications of biobased packaging materials for the food industry. Food Addit. Contam. 19, 172–177 .

Weng W, Zheng H. 2015. Effect of transglutaminase on properties of tilapia scale gelatin films incorporated with soy protein isolate. Food Chem. 169, 255-260.

Wittaya T. 2012. Protein-based edible films: Characteristics and improvement of properties, in: Structure and Function of Food Engineering, Ayman Amer Eissa (Ed.),pp. 159-181, IntechOpen.

Ye M, Neetoo H, Chen H. 2008. Control of Listeria monocytogenes on ham steaks by antimicrobials incorporated into chitosan-coated plastic films. Food Microbiol., 25(2), 260-268.

Zahiruddin SMM, Othman Siti, TawakkalIntan, A Talib, Rosnita. 2018. Mechanical and thermal properties of tapioca starch films plasticized with glycerol and sorbitol. Food Res. 3. 160-166.

Zhang N, Xu J, Gao X, Fu X, Zheng D. 2017. Factors affecting water resistance of alginate/ gellan blend films on paper cups for hot drinks. Carbohydr. Polym., 156, 435-442.

Zhang W, Li X, Jiang W. 2019. Development of antioxidant chitosan film with banana peels extract and its application as coating in maintaining the storage quality of apple. 1205-1214 Int. J. Biol.Macromol.

Zhong Y, Godwin P, Jin Y, Xiao H. 2020. Biodegradable polymers and green-based antimicrobial packaging materials: A mini-review. Advanced Industrial and Engineering Polymer Research, 3(1), pp.27-35.

Zhou FF. 2014. The Theory of Green Packaging Design and its Application. Applied Mechanics and Materials, 635–637, 248–252.

13

Processing of Banana Flower to Value Added Products: A Review

Pooja Jha, Murlidhar Meghwal* **and** ***Pramod K. Prabhakar***

ABSTRACT

Banana plants areextensively known for their fruits butby-products (pseudo-stem, root, leave, male banana flower, and banana peel) from banana farms also have potential applicability because recycleprocess of these wastes can convert waste of banana farms into an outstandingresource of valuable raw materials for other industries. This practice avoids huge losses of unused biomass and environmental hazards. Banana flowers are discarded in banana farms during the harvesting of banana fruit bunches and produce a great amount of agricultural waste. Consumption of banana flower as a vegetable is trendy in some parts of Asian countries but the cleaning process of banana flower is an unwieldy practice because of the time and the laborengaged in separating the calyx and the pistil portion of these flowersbefore processing. Itis an abundant source of well-balanced essential amino acids, unsaturated fatty acids, dietaryfiber, vitamins, minerals,and flavonoids. This flower also having tremendous medicinal properties as they possess several secondary metabolites. Thepresent review discusses the transformation of this agricultural waste into a different value-added product. The transformation of banana flowers into highly commercial outputs can be beneficial forsupplementary income to small scale farming industries.

Keywords: Banana, Banana flower, By-product, Value-added product, Antioxidant, Pre-treatments, Processing.

13.1 Introduction

Agricultural waste comprises mostly under-exploited residuesfromthe growing and processing of raw agriculturalproducts such as fruits, vegetables, crops, poultry, meat, and dairy products. A huge quantity of agricultural and food waste is produced all over the world with the evolution of technology and population. These agricultural wastes are produced from several agricultural

*Corresponding Author

bustles include farming, harvesting, livestockproduction, and aquaculture. Earlier, agriculture and food wastes were either thrown away and burnt off or let rot in farmshowever this practice is harmful to the environment. Consequently, utilization and processing of this waste require to be extended to facilitate the sustainable exploitation of agricultural waste and minimize environmentalcontamination. Recently agriculture waste valorization practices have been focused on by researchers becausethese wastes have extensive applicationsas a result of their low price and easyavailability and reusability. Agriculture and food waste has been espoused as cheap resources for new functional and nutraceutical products or vital phytochemicals with numerousmedicinal properties.

Banana is a luscious and very well-liked fruit all over the world. Bananas are the most extensively grown fruit crops in India (Rajoriya *et al.*, 2018) and the benefit of this fruit is its availability throughout the year. India is a top producer of bananas followed by the country China and the Philippines. About 1200 cultivars of bananas are known in the world (Florent *et al.*, 2015). Banana is chiefly cultivated for ripe fruits, unripe fruits, and leaves. Each banana tree produces a bunch of fruit only once in its life span. After which, it becomes waste, and hence it cut down. Due to this large amount of waste is generated from a banana crop in one crop season which includes stem, leaves, banana male buds, peel, and roots. Presently, a huge quantity of waste is dumped in our country as waste and farmers are facing massive difficulties in arranging the accrued banana waste. Thus, avaluable economic means of minimizing these environmental issues by recycling these waste such as extraction of fiber and manufacturing of many food products, food supplements, functional food, nutraceutical products, fertilizer, bio-chemicals, papers, etc. from banana waste can be recommended.

The banana plant contains both male and female flowers which come out of the plant separately. The first 5-15 basal nodes generate female flowers and the upper digital nodes turn outinto male flowers. The female flowers develop into fruits and male flowers are thrown away in large quantities during harvesting of banana bunches and produce significant quantities of agriculture waste with very low economical value.It is typically red or purple-red color, oval-shaped, and connected to the end of the bananabunches. The flower constitutes several parts such as bract, calyx, pistils, and white flowers. Banana flowers contain lots of small whitish flowers, which are packed in a purple or red color bract. Whitish flowers contain calyx and pistils which are generally removed before preparation of food dishes because pistils are imparting bitterness to food products. Each part of the banana flowers is shown in Figure 1.

Banana flower Purple Bract Whitish Flower Calyx Pistil

Fig. 1: Different parts of the banana flower

In recent times, it is a great challenge for farmers to dispose of it. Now it's become a topic of great interest for the researchers to exploit these by-products, indifferent food and non-food application such as flavor, natural coloring material, micronutrient supplements, functional and nutraceutical products, extraction of fibers and phytochemicals, animal feeds, and bio-fertilization (Padam *et al.*, 2014).

13.2 Consumption of Banana Flowers

Consumption of banana flower as a vegetable is trendy among people living inSouth Asian and Southeast Asian countries. Both the heart and fleshy part of the bracts are edible. Banana flower is cooked to prepare different forms of cuisines. Usually,the outer bracts, calyx, and pistil portion of flowers are removed before cooking. Banana flowers are consumed as a vegetable, either raw as a salad or steamed in the form of soups, curries, and fried foods. Different food products, for example, dehydrated banana flower, banana flower powder, pickles, canned products, and herbal tea prepared from this flower are also well known (Wickramarachchi and Ranamukhaarachchi, 2005).

13.3. Nutritional Composition of Banana Flower

Banana flowers have tremendous nutritional values because these flowers are a storehouse of high- quality protein due to their well balanced essential amino acid, unsaturated fatty acids, dietary fiber, vitamin's A,C, and E, minerals,for example, potassium, magnesium, iron, copper, and flavonoids (Sheng *et al.*, 2010; Jahan *et al.*, 2010; Bhasker *et al.*, 2011, Swe, 2012; Florent *et al.*, 2015). Major amino acids present in banana flowers are glycine, leucine, alanine, and aspartic acid. Banana flowers have a large quantity of unsaturated fatty acids especially linoleic and gamma-linolenic acids; they could be exploited to avoid atheroma formation that leads to arteriosclerosis

(Florent *et al.*, 2015). These flowers are a rich source of dietary fibers and are helpful for the control of obesity and diabetes. Dietary fiber is helpful to control the cholesterol level in the human body. It is advised that an adult person should take 20-30 grams of dietary fiber per day to be healthy. The deficiency of dietary fiber in the human body can cause diverticulosis, risk of colon cancer, breast cancer, and constipation. Banana flowers contain ample potassium which facilitates to balance of sodium- potassium levels in the human body and thus helps in the reduction of hypertension or high blood pressure. Hypertension and erectile dysfunction are strongly linked together. As potassium helpful to decrease the risk of high blood pressure, it also helps in healing erectile dysfunction. This flower decreases the risk of stroke. Iron present in banana flowers facilitates boosting the production of hemoglobin which is helpful for the patient who is suffering from anemia. Banana flowers that are storage houses of many nutrients, vitamins, and minerals can be used for infantile malnutrition and the weak human body.

13.4 Antioxidant Activity of Banana Flowers

Antioxidants that scavenge free radicals play a significant role for medical reasons. Synthetic antioxidants buthylhyroxyanisole (BHA) and butylhydroxtoulene (BHT) have several side effects. To avoid the side effects of synthetic antioxidant alternative antioxidants from plant sources can be a great option to replace synthetic antioxidants. Thus, there is currently great attention placed onthe search for alternative antioxidants from natural sources. Extract prepared from banana flowers is also believed to as effective sources of natural antioxidants include phenolic and flavonoids. These antioxidants neutralize the free radicals formed as a result of various metabolic processes in the body and can be used aspostprandial hyperglycemia regulators (Marikkar *et al.*, 2016). If the free radicals are not neutralized, their unstable electrons reactwith the DNA and proteins of human cells and alter their properties. This canlead to several chronic conditions, including cancer and heart disease. Antioxidants help to enhance human resistance to various diseases.

13.5 Bio-Active Phytochemicals Present in Banana Flower

Banana flowers are used in traditional medicine fora hundred years to alleviate different diseases and health problems.Bioactive compounds present in the banana flower might be accountable for the beneficial effect on human health. Extract prepared from banana flowers possesses different bioactive compounds such as alkaloids, glycosides, steroids, saponins, tannins, flavonoids, and terpenoids which have several health benefits (Swe, 2012; Timsina and Nadumane, 2014). These bioactive constituents were identified to hold a range of bioactivities for example the antioxidant, antimicrobial, antivirus, and anticancer. These bioactive compounds havethe potential to defeat health problems, banana flower-based products can be produced in industries as botanical drugs, dietary supplements, and functional foods. Saponins are helpful to lower bad cholesterol, boost our immunity against infection, and also inhibit the growth of cancer cells. They also have antioxidant activity and so can reduce our risk of chronic diseases such as cardiovascular disease. Banana flowers are also an excellent source of flavonoids. These phytochemicals found in many plant-based foods help prevent damage to DNA cells by neutralizing free radicals. They also help lower cholesterol, are anti- inflammatory, anti-cancer, and anti-aging. In recent times, most researchers havefocused to explore bioactive secondary metabolites to find outnovel plant drugs.

13.6 Pharmacological Activity of Banana Flower

Since ancient times, banana flowers were used to treat many different types of diseases. A banana plant is not only a food crop but it is also known for medicinal uses. Apart from the banana fruit, other parts of the banana tree are known to have medicinal properties beneficial to mankind. There is no scientific information abouttheside effects and drug interactions of banana flower. Many research works conducted in vivo and in vitro showed the health benefits of the consumption of banana flowers (Bhaskar *et al.*, 2011; China *et al.*, 2011; Singh, 2017). Extract of the banana flowers has anticancer activity, antidiabetic activity, antioxidant activity, antimicrobial activity (Pushpaveni *et al.*, 2019). Traditionally banana flowers are consumed to cure certain diseasesfor example heart pain, diarrhea, asthma, and stomach cramps (Sumathy *et al.*, 2011). According to Kumar *et al.*, (2012), it can be used for the treatment of bronchitis,dysentery, ulcers, etc. Banana flowers do exhibit some significant antimicrobial activity against a few tested microorganisms Extract of banana flowers can inhibit the growth of some pathogenic bacteria and may help out to cure wounds and to avoid infections (Mumtaz *et al.*, 2009; Sumathy *et al.*, 2011). According to Kumar (2012), the banana flower has tremendous nutritional value and health benefits such as for the treatment of bronchitis, dysentery, ulcers, etc. Banana flowers are extensively used in traditional drugs for the healing of diabetes mellitus. Pari and Umamaheswari, 2000 also reported that banana flowers are having antihyperglycemic activity and help in the diminution of blood sugar level and increased hemoglobin in the blood. Consumption of cooked banana flower with card or yogurt is the most effective treatment in case of excessive and painful menstruation bleeding. The combination of curd and cooked banana flowers enhances the level of progesterone in the human body and in that way decreases bleeding due to menorrhagia.This flower also helpful to increase the milk of lactating women and treat dysmenorrhea.

13.7 Pre-Treatment of Banana Flower

Banana flowers are highly susceptible to enzymatic browning while exposes to atmospheric conditions. Processing steps for example cutting, slicing and dehydration always endorse browning which leads to reduce the sensory quality of end products. Development of brown pigments occurredas the polyphenol oxidase enzyme reacts with phenolic compounds. Developing a process to avoid enzymatic browning is significant for the processing of banana flowers due to increasing consumer acceptability. To avoid this enzymatic browning and to eliminate the bitterness and starchy flavor the banana flowers are usually pre-treated before processing. Following pre-treatment is most commonly used before processing. Hot water blanching is generally used for reducing enzymatic browning in fruits and vegetables. But this practice is found ineffective for preserving the banana flower because of enzymatic browning which reduces market demand for the processed product (Wickramarachchi and Ranamukhaarachchi, 2005). Soaking banana flowers in buttermilk, citric acid solution, lemon juice solution, salt solution, rice rinse water before processing improves the flavor and color of end products (John *et al.*, 2014; Tasnim *et al.*, 2020). The citric acid solution performs as a natural antioxidant and avoids the development of brown pigments which is caused by a reaction of PPO with phenolic compounds. The PPO enzyme produces o-quinones which breakdown into non-enzymatic oxidative polymerization leading to the development of brown pigments. Thus, to minimize the formation of brown pigment, it works as antioxidants and reduces enzymatic browning reaction take place in banana flowers. Dipping sliced banana flowers in a solution containing 0.2% citric acid for 30 minutes before different processing such as cooking, drying, etc of banana flowers can arrest browning reaction (Ambeors *et al.*, 2018).

13.8 Processing of Banana Flower

The shelf life of banana flowers is too short because it contains a high amount of moisture and also highlysusceptible to enzymatic browning. Under an open atmosphere, these flowers could remain fresh for a few hours only when kept because they are more susceptible to enzymatic browning. Banana flowersbloom continuously drops the petals at ambient temperatures while it starts rotting athigher temperatures and chilling temperaturecauses blackening in flowers. Banana flowers having high nutritional and medicinal values still,the consumption of banana flowers maybe not popular due to the unwieldy preparation process. Developing a preserved product from the banana flowers not only will increase the shelf life of the product but also will increase consumer's demands too.There are different types of processed banana flowers as described below:

13.8.1 Minimal Processing

Nowadays consumers are giving preference to convenience food due to busy lifestyles. A busy lifestyle of urban people and cumbersome cleaning procedure of the banana flowers diminishing the consumption of banana flowers.Cleaning of the banana flowers is a burdensome procedure because it takes too much time and the hard work involved in removing the calyx and pistils from each flower before processing. Minimally processed vegetables are gaining popularity among urban people. Freshly processed banana flowers in a ready-to-cook form will satisfy consumer's need for time and healthy food. Minimal processing of fresh banana flowers could be employed to sustain the fresh-like quality characteristics and also offer the convenience demand of the consumers.Minimal processing assists in the rapid and simple preparation of ready meals. The increasing demand for freshly processed food with the assurance of safety has recommended those in research to focus on studying new ways of increasing the shelf life of fresh-cut products. The shelf life of minimally processed banana flowers depends on pre-treatments, storage conditions, and packaging materials. Physiological Weight Loss (PWL) affects the quality of minimally processed banana flowers and is themainreason for deterioration during storage. PWL of minimally processed banana flowerswas observed in a slower phase of a pre-treated sample with 0.2% citric acid solution refrigerated condition (Ambeors *et al.*,2018). Banana flowers could stay fresh for 3 days in non-ventilated polypropylene (40-micron thickness) packed bag under ambient as well as refrigerated storage (Ambeors *et al.*, 2018).

13.8.2 Dehydrated Banana Flower

To increase the shelf life of the banana flowers, they are usually dehydrated. Drying is one of the best practices to produce shelf-stable and high-quality products. The drying technique prevents the growth of microorganisms and other deteriorative reactions as it removes moisture content. Drying persuades a significant reduction in weight and volume reduces packing, storage, and transportation costs, and also tofacilitates the product to be stored under ambient conditions.Dehydrated banana flowers are also having great nutritional value based on their high content of potassium and fiber.The dried flower scan be rehydrated and used in ready-to-eat foods (John *et al.*, 2014).

Fig. 2: Image of marketed dried banana flower

13.8.3 Banana Flower Powder

Considering the high nutritional value of the banana flower, this flower can be incorporated into thediet as dehydrated powder and consequently effortlessly incorporated into food (Fathima *et al.*, 2018). Banana flower powders also having a great nutritional supplement based on their high dietary fiber content, essential amino acids, unsaturated fatty acids, minerals, vitamins, and antioxidant properties (Arya Krishnan and Sinija, 2016). Banana flower powder can be easily incorporated for the preparation of different food formulations such as bakery products (biscuits, cakes, cookies, Hummus), traditional dishes (Anand and Sharma, 2019; Tasnim *et al.*, 2020). Processing steps for preparation of banana flower powder is given in Fig. 4

Fig. 3: Image of marketed dried banana flower

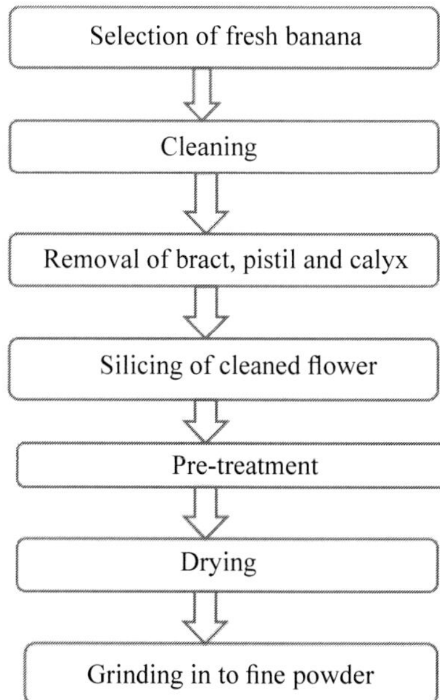

Fig. 4: Processing steps for production of banana flower powder.

Fig. 5: Image of marketed banana flower pickle

13.8.4 Banana Flower Pickle (Thokku)

The banana male budsare waste material of banana farms and having less economic value. It is transformed into high value-added products by making flower pickles. The process of making pickles from banana flowers includesthe removal of the pistil, blanching, grinding, and addition of spices and oil. The products are tasty and can be stored for a year at room temperature. The product is suitable for all age groups. This technology can be adaptable to banana growing farmers and for small scale industries.

13.8.5 Canned Banana Flowers

Canned Banana flowers are convenient to use where fresh banana flowers are not available. Banana flowers in brine can be used for the preparation of many traditional Thai dishes like curry or served as a vegetable with spicy Thai chili pastes.

Fig. 6: Image of marketed canned banana flower

13.8.6 Banana Flower Tea

Banana flowers can be converted into a caffeine-free herbal tea with a pleasant flavor and high nutritional value.Banana flower tea is prepared using steps including cutting or shredding, drying roasted under certain conditions. This herbal tea is rich in antioxidants to boost immunity and reduces anxiety.

13.8.7 Banana Flower Oil

Banana flower oil is rich in Vitamin E and effective against skin alterations and the aging process because vitamin E activates microcirculation of the skin. Vitamin E acts asa cell antioxidant, an anti- inflammatory agent that helps in delaying cell aging, enhances the elasticity of the skin, protects against UV damage, improves blood microcirculation,and reduces wrinkles on the skin. Oil extracted from banana flowers can be used for the preparation of face creams, gels, and body lotion for skin disorders, skin softness, and anti-aging care. Banana flower oil can be stored up to 18 months in a tightly closed container, at a temperature between + 5°C to 12° C, away from light, heat, and humidity sources.

Fig. 7: Image of marketed banana flower tea

13.9 Challenges and Future Scenario

Safety criteria should be the main concern while selecting edible flowers for therapeutic purposes assomeedible flowers used as medicine are poisonous. Consumption of banana flower as a vegetable and traditional medicine is popular among Asian countries but it is essential to identify the toxicity level to avoid overdoses or poisoning effect on the human body. Banana flowers are having antioxidant, antimicrobial, antidiabetic, anticancer property which can be utilized in the future to prepare a generic drug by Nano-technology.

13.10 Conclusion

Banana flowers are having a unique nutritional value so that it is believeda potent sourceof food supplement for health benefits. Consumers always preferconvenience foods that are delicious and have high nutritional value with health benefits. Processing of banana flowers possibly will provide additional advantages in minimizing the banana waste and growing the use in food and nutraceutical application. Maximum utilization and incorporation of these banana flowers in food and nutraceutical products can be executedfor value addition because of its exceptional nutritional and pharmacological values of banana flowers.Recently, more concentration has been centered on the utilization of agricultural by-products. Therefore, the utilization of banana flowers could provide supplementary benefits in minimizingwaste and increasing the use in food science.

13.11 References

Ambrose DCP, Sumithra V, Vijay K, Vinodhini K. 2018. Techniques to Improve the Shelf Life of Freshly Harvested Banana Blossoms. Curr. Agri. Res., 6, 141.

Anand S, Sharma M. 2019. Product Development from Banana Blossom Powder and Indian Gooseberry Powder for Anaemic Adolescent. Int. J. Health. Sci. Res., 9, 273.

Arya KS, Sinija VR. 2016. Proximate Composition and Antioxidant Activity of Banana Blossom of Two Cultivars in India. Int. J. Agri Food Sci Tech., 7, 13.

Florent AW, Loha AMB, Thomas HE. 2 0 1 5 . Nutritive Value of three varieties of banana and plantain blossoms from Cameroon. Greener J. Agric. Sci., 5, 052.

Bhaskar JJ, Shobha MS, Sambaiah K and Salimath PV. 2011. Beneficial effects of banana (Musa sp. var. elakki bale) flower and pseudostem on hyperglycemia and advanced glycation end- products (AGEs) in streptozotocin-induced diabetic rats. J. Physiol. Biochem., 67, 415.

China R, Dutta S, Sen S, Chakrabarti R, Bhowmik D, Ghosh S and Dhar P. 2011. *In vitro* antioxidant activity of different cultivars of banana flower (Musa paradicicus L.) extracts available in India. J. Food Sci., 76, 1292.

Fathima ZP, Vijayalakshmi D, Suvarna VC, Yatnatti S. 2018. Processing of Banana Blossom and its Application in Food Product. Int.J.Curr.Microbiol.App.Sci., 7, 1243.

Jahan M, Warsi MK, Khatoon F. 2 0 1 0 . Concentration antimicrobial activity of banana blossom extract incorporated chitosan-polyethylene glycol (CS-PEG) blended film, J. Chem. Pharmaceut. Res. 2, 373.

John SG, Sangamithra A, Veerapandian C, Sasikala S, Sanju V. 2014. Mathematical Modelling of the Thin Layer Drying of Banana Blossoms. J Nutr Health Food Eng, 1(2), 00008.

Kumar S, Kumar V, Rana M, Kumar D. 2012. Enzyme inhibitors from plants: an alternate approach to treat diabetes. Phcog. Commn., 2, 18.

Marikkar JMN, Tan SJ, Salleh A, Azrina A, Shukri MAM. 2016. Evaluation of banana (*Musa* sp.) flowers of selected varieties for their antioxidative and anti-hyperglycemicpotentials. Int. Food Res. J., 23(5), 1988-1995.

Mumtaz J, Warsi M, Khatoon F. 2009. Concentration influence on antimicrobial activity of banana blossom extract-incorporated chitosan-polyethylene glycol (CS-PEG) blended film. J. Chem. Pharm. Res., 2, 373.

Padam BS, Tin HS, Chye FY, Abdullah MI. 2014. Banana by-products: an under-utilized renewable food biomass with great potential. J. Food Sci. Tech. 51, 3527.

Pari L, Umamaheswari J. 2000. Antihyperglycaemic activity of Musa sapientum flowers: effect on lipid peroxidation in alloxan diabetic rats. Phytother. Res., 14, 136.

Pushpaveni C, Visagaperumal D, Chandy V. 2019. A review on banana blossom: the heart of banana. World J. Pharm. Res. 8, 440.

Rajoriya P, Singh VK, Jaiswal N, Lall R. 2018. Optimizing the effect of plant growth regulators on in vitro micro propagation of Indian red banana (*Musa acuminata*). J. Pharmacogn. Phytochem., 7, 628.

Sheng ZW, Ma WH, Jin ZQ, Bi, Y, Sun, ZG, Dou HT, Li JY, Han LN. 2010. Investigation of dietary fiber, protein, vitamin E and other nutritional compounds of banana flower of two cultivars grown in China. Afr. J. Biotechnol., 9, 3888.

Singh S. 2017. Banana blossom-an understated food with high functional benefits. Int. J. Curr. Res..9, 44516.

Sumathy V, Lachumy SJ, Zakaria Z, Sasidharan S. 2011. *In Vitro* Bioactivity and Phytochemical Screening of *Musa Acuminata* Flower. Pharmacologyonline, 2, 118.

Swe KNN. 2012. Study on Phytochemicals and Nutritional Composition Of Banana Flowers of Two Cultivars (*Pheekyan* and *Thee hmwe*). Universities Res. J., 5.

Tasnim T, Das PC, Begum AA, Nupur AH, Mazumder AR. 2020. Nutritional, textural and sensory quality of plain cake enriched with rice rinsed water treated banana blossom flour. J. Agri. Food Res., 2,100071.

Timsina B, Nadumane VK. 2014. Anti-cancer potential of banana flower extract: An in vitro study. Bangladesh J Pharmacol. 9, 628.

Wickramarachchi KS, Ranamukhaarachchi SL. 2005. Preservation of Fiber-Rich Banana Blossom as a Dehydrated Vegetable. ScienceAsia, 31, 265.

14

Carbon Foot Printing and Green Technology: An Environment Friendly Approach in Dairy Sector

Partha Pratim Debnath*, ***Syed Mansha Rafiq*** **and** ***Nandini Dutta***

ABSTRACT

The rapid growth in population has created the need of industrialization and as a result there has been a significant increase in milk production and processing. Milk is a commodity which is consumed by persons of all age groups due to its nutritive value and health promoting attributes. However, the rapid growth in dairy sector has adversely affected the environment by emitting various greenhouse gases and thereby, resulting in global warming. Apart from it, there has been an increased usage of non- renewable resources like natural gas and coal in dairy sector. As a result, a viable balance has to be maintained between milk production and processing, environmental health and the rapidly growing population. This can only be achieved with the application of green or environment friendly technologies in all the stages of milk production and processing. This chapter will mainly deal with the various sources of greenhouse gases from dairy sector and on application of life cycle assessment and carbon foot printing for computing the impact of greenhouse gases on the environment. Also, this chapter will enclose various green technologies which can be applied in dairy sector to curb the greenhouse gas emissions.

Keywords: Carbon footprint, LCA, Greenhouse gases, Green Technology, Milk production, Milk Processing

14.1 Introduction

Milk is considered as almost a complete food. Due to its, nutritive value and wide health promoting attributes, the demand of milk and milk products is increasing progressively. As a result, dairy sector plays a significant role in India's economy. Dairying is also a crucial income source for many lakh rural

*Corresponding Author

farmers and it is also serving the part of providing employment opportunities. As per Federation of Indian Chambers of Commerce and Industry (FICCI, 2020), the dairy and animal husbandry sector combined, contributes almost 4.2% of India's GDP (Gross Domestic Product), thus providing livelihood to approx seventy million rural households. In 2017-18, the output value of the India's agriculture and allied sector stood at Rs. 28 lakh crore, of which the dairy sector alone constituted almost Rs. 7 lakh crore, therefore dairy sector accounts for more than 25% of total output value (National Accounts Statistics, 2019). This is mainly because of the tremendous growth (about 6.5%) in milk production from 176.3 million tonnes in 2017-18 to 187.7 million tonnes in 2018-19 with a per capita availability of 394 grams milk/ day (National Dairy Development Board, 2020). With 21% of global milk production (827.88 million tonnes), India holds the first spot in the world milk production. India also has the largest bovine population in the world.

Although, this huge milk production and large livestock population, boosts the economy and provides employment opportunities, but this success is at the expense of environmental pollution. The dairy sector is being held accountable for emitting greenhouse gases (mainly, carbon dioxide, methane and nitrous oxide), which elevates the earth's surface temperature and deteriorates the global climate. In 2010, the greenhouse gas emission from India has been reported as 2.136 billion tonnes in CO^2 equivalent, of which, the maximum portion (71%) is being contributed by the energy sector, whereas, the agriculture sector, industrial product and process sector and waste sector accounts for 18, 8 and 3%,respectively. The dairy sector contributes almost 58% of the greenhouse gas emissions from the agriculture sector (Indian Dairy Association, 45th Dairy Industry conference). Although, India contributes 21% of the world milk production, but it only accounts 14.1% of greenhouse gas emissions. The dairy sector is also responsible for overusing the land and water resources. In addition to this, recently consumersare looking for organic milk, in which preservatives, antibiotics or any other chemicals should be absent. Therefore, environment friendly milk handling practices are the need of hour, for maintaining a balance between healthy environment, sustainable milk production and processing and a healthy population.

Green or eco-friendly technology is themethod to reduce the depletion of natural resources by saving energy, decrease the environmental pollution by disposing the waste in eco-friendly manner and minimize the product wastage by utilizing the by-products generated. Green technology should provide sufficient food for the rapidly growing population while decreasing the environmental impact. As a result, sustainable development is a necessity, i.e. development which fulfills the present needs, without damaging the future generation's ability to satisfy their needs. This chapter will mainly focus upon the carbon footprint of milk production and processing through life cycle assessment (LCA) tool. Also, this chapter will deal with the application of green technologies in milk production and processing (including packaging) such as to reduce the carbon foot print.

14.2 Life Cycle Assessment (LCA) of the Dairy Sector

LCA is a cradle to grave or cradle to cradle method which is used to evaluate the environmental affect related with the complete life cycle of a product from acquisitionof raw materials to transportation and processing, distribution, use and disposal of the product. The International Organization for standardization (ISO 14040:2006), defines LCA as the accumulation and assessment of inputs, outputs and environmental effect of the entire life cycle of the product. However, LCA is being termed as carbon footprint analysis, if the LCA is only estimating the contribution of any product to global warming (International Dairy Federation, 2015).

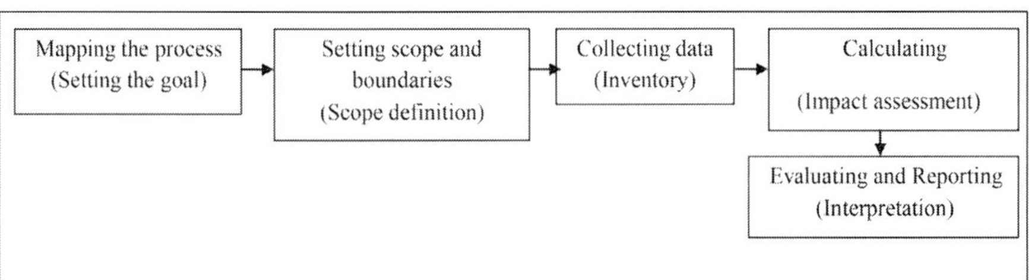

Fig. 1: Flow Diagram to Conduct LCA (adapted from IDF, 2015)

14.2.1 Goal

The goal is to evaluate the carbon footprint of milk. For evaluating the carbon footprint of dairy or cattle farm, one kg of FPCM (Fat and protein corrected milk) is being chosen as functional unit. The formula for calculating the functional unit in dairy farm is given below— FPCM (Kg/ yr) = Milk production (kg/yr) * [0.1226*Milk Fat% + 0.0776* True Protein% + 0.2534] (IDF, 2015).

But at the milk processing factory, where the milk is packed, one kg of product with a% fat and b% protein is being considered as the functional unit.

14.2.2 Scope and System Boundary

This step includes all the unit processes during the product manufacture. Therefore, this category involves the analysis of all unit operations in dairy farm (animal husbandry, management of manure, production of feed and energy utilized) and in the processing industry (raw milk transportation from village co-operative societies to the processing plant, processing of milk,packaging, waste products generated and its disposal and transportation of processed products to the retail shops and consumers). All the sources which can emit greenhouse gases should be considered here. Allocation is required if the by-products of crop (e.x.: soy meal is obtained as a by-product during production of soy oil from soybeans) and milk (e.x.: meat is derived as a co-product), are the parts of milk production system. Allocation is also necessary to divide the environmental load between milk and different dairy products processed in the same industry like powder, butter, ice cream etc.

14.2.3 Life Cycle Inventory (LCI)

This step includes the collection of all primary (collected) and secondary (database, article, report) data related to all the unit operations within the system boundary (i.e. from handling animals at the farm to the delivery of the finished milk or milk product to the consumer). Also, from which source the data is being collected should also be specified. Most of the farm related data's can be collected from NDDB's Information Network for Animal Productivity and Health (INAPH),Department of Animal Husbandry and Dairying (DADF) and by personally communicating with field veterinary officers, farmers and Krishi Vigyan Kendras (KVKs). However, most of milk processing data's can be collected from village co-operative societies (VCS) and from co-operative milk unions (for example, AMUL) present in each state.

14.2.4 Life Cycle Impact Assessment (LCIA)

In this step, the total impact on climate by the entire dairy sector is being summarized. The greenhouse gas emissions of all the unit operations are being summed up and expressed in CO_2- equivalent ,by multiplying the individual greenhouse gas emissions with the appropriate global warming potential (GWP) factor { for example, GWP of carbon dioxide, methane and nitrous oxide are 1, 34 and 298 respectively} (Garg *et al.*, 2018). The GWP factors are obtained from Intergovernmental Panel on Climate Change (IPCC).

14.2.5 Evaluating and Reporting

The LCA report should positively have a section which includes all the preventive measures, that has to be considered to reduce the greenhouse gas emissions. The report should also be relevant to the company or user and it should include all the sources of emissions within the system boundary (any exclusion should be disclosed and justified).

14.3 LCA of Milk Production

The greenhouse gas (GHG) emissions from dairy farm mainly include those emissions which are emitted from production of feed, enteric fermentation, management of manure and soil and combustion of fuel (Table 1). Broadly, the emissions generated can be categorized into primary and secondary sources. Emissions originating from the dairy farm during the actual milk production system can be defined as primary sources, while secondary sources are derived during the manufacturing of resources utilized in the milk production process. Carbon dioxide (CO_2), methane (CH4) and nitrous oxide (N_2O) are the three most important GHG which are emitted during the milk production. CO_2 is mostly emitted by the farm animals (cow, buffalo etc.) during respiration. However, the feed plants or crops consume CO_2 and emits O_2 during respiration, thus the plant or crop biomass remaining (after the crop is harvested or after the farm lands are grazed by animals), is a rich source of CO_2, which can get released in the environment if the plant biomass is rapidly burnt. Manures and barn floors emits negligible amounts of CO_2. Enteric fermentation occurs during digestion of feed by the farm animals and it is the main component responsible forthe release of methane. Methane gets released from the animals either by belching or by respiration. Enteric fermentation is influenced by the feed intake, particularly the starch content in the feed. Manures which have been stored also play a part in methane emissions, whereas the manures which has been deposited by the animals in bran or on the grazing land, performs a minor role in methane emission. Nitrous oxide mostly gets emitted by the denitrification and nitrification processes, where the feed plants or crops are cultivated. This is mainly because of the increased usage of fertilizers in soil. Nitrous oxide also gets emitted if the manure is stacked for quite a long time. Apart from the above discussed primary sources, secondary emission sources include manufacturing of fuel, fertilizer, electricity, pesticide, machinery and plastic, which are utilized in feed production, animal maintenance and manure handling. Among all the emission sources, enteric fermentation is the major contributor of greenhouse gases.

Table 1: GHG Emissions during Milk Production in Indian Dairy Sector in 2015

Source of GHG emission	Quantity of GHG emitted (mt CO_2- eq/year)
Feed production (CO_2)	23.5
Feed production (CH_4)	28.9
Feed production (N_2O)	13.2
Total GHG from feed production	65.6
Enteric fermentation (CH_4)	457.4
Manure management (CH_4)	37.4
Manure management (N_2O)	25.6

Source of GHG emission	Quantity of GHG emitted (mt CO_2- eq/year)
Managed soils (N_2O)	22.5
Fuel combustion (CH4, N2O)	25.6
Post-farm gate (CO_2,CH_4, N_2O,Hydroflurocarbons)	11.3
Gross total GHG emissions	663.7
Net total GHG emissions	449.8

(Garg *et al.,* 2018)

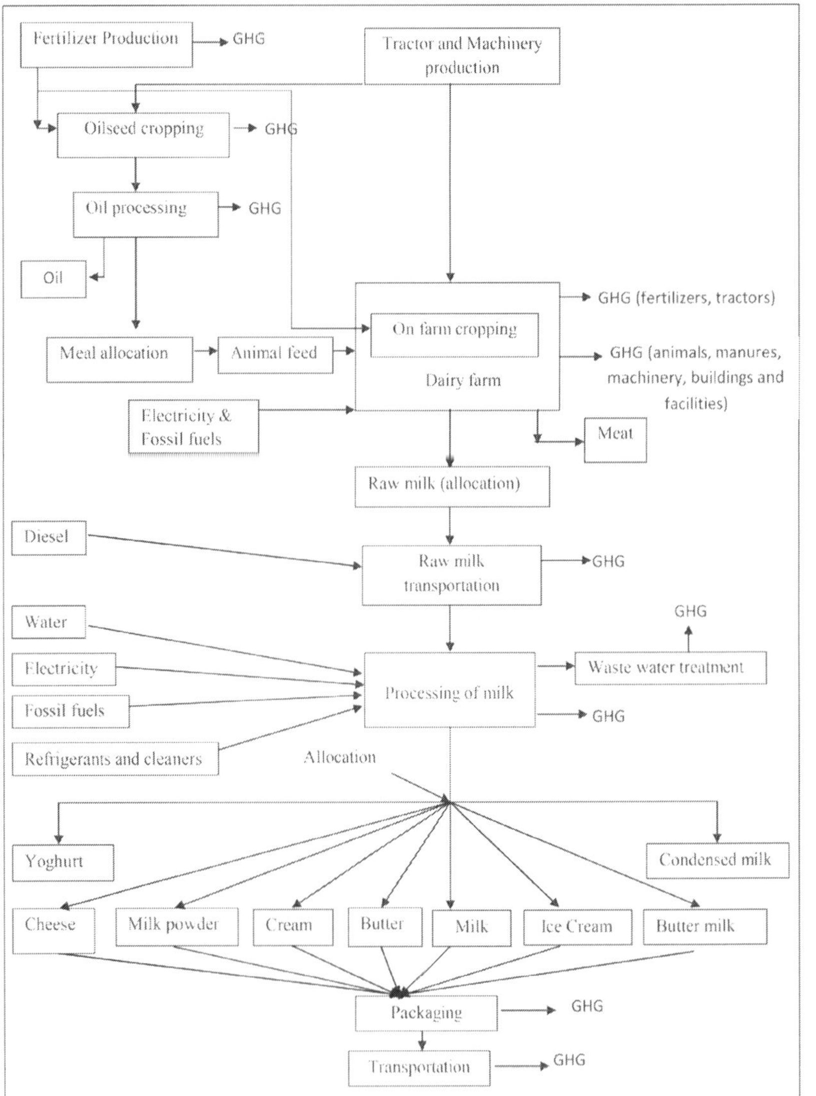

Fig. 2: Carbon foot printing of milk production and processing system (Verge *et al.*, 2013)

14.4 LCA of Milk Processing

Milk is a highly perishable product. As a result, milk has to be immediately processed or refrigerated, if the milk is to be processed later or transported over long distances. In most of the dairy industries in India, raw milk after being collected in village co-operative societies is immediately chilled to below 4°C and is being transported by roadways or railways to processing plant with the help of refrigerated or insulated milk tankers. The fuel used for transporting the milk, plays a part in greenhouse gas emissions.

A singledairy industry not only manufactures fluid milk, but also produces cream, butter, dahi, ice cream, paneer, cheese, evaporated and condensed milk, skim milk powder and many others. Heat treatment (hot water/ steam), cooling (refrigerants), evaporation, drying and cleaning are basically performed in almost all dairy industries. These five treatments are responsible for 50% of energy consumption in fluid milk production and 96% in milk powder production (Ramirez *et al.*, 2006). Dairy plants manufacturing condensed milk and powder consumes higher amount of energy as compared to fluid milk processing plant (Figure 3). Energy in dairy industry is mostly derived from electricity followed by natural gas and coal. These energy sources play a part in greenhouse gas emissions. Also, different dairy products have varied environmental impacts or different green house gas emissions (Table 2) because of their differences in production, processing and distribution (for example, ice cream is stored at freezing temperatures below -18°C whereas toned milk is stored at refrigerated temperatures below 4°C) (Arcand *et al.*, 2012).

Table 2: Carbon footprints of different non-packaged dairy products

Dairy product	Amount (kg of CO_2 equivalents (CO_2e)/ kg of product)
Cheese	5.3
Creams	2.1
Sour cream	2.5
Yoghurt	1.5
Fluid milk	1.0
Butter milk	1.1
Frozen dairy products	2.1
Milk powder	10.1
Concentrated milk	3.1
Butter	7.3

(Verge *et al.*, 2013)

Apart from energy, significant amount of water is being utilized in dairy industry, which is an alarming concern considering the fresh water scarcity. Water is used as an ingredient in batch formulation for preparation of products like toned milk, double toned milk etc. Water is also utilised in processing, washing of trucks, cleaning of crates and CIP (clean in place) of process lines, generation of steam and in cooling tower and also for drinking purposes in the plant (Figure 3). The consumption of water for fluid milk production ranges from 0.5-4 liter/ liter of milk, whereas for milk powder manufacture, water usage is only 1.5 liter/ liter of milk (Rad and Lewis, 2014).

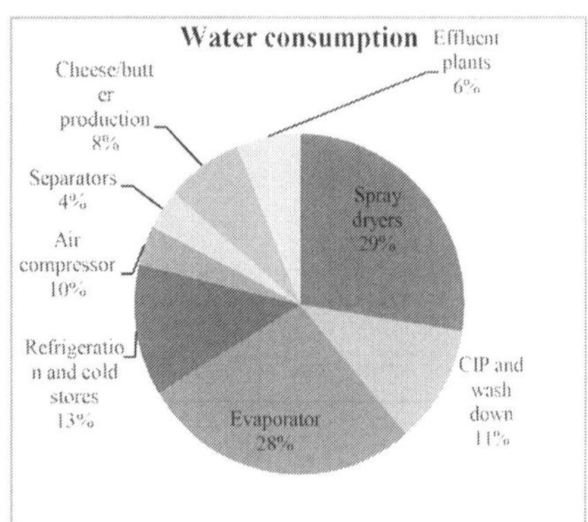

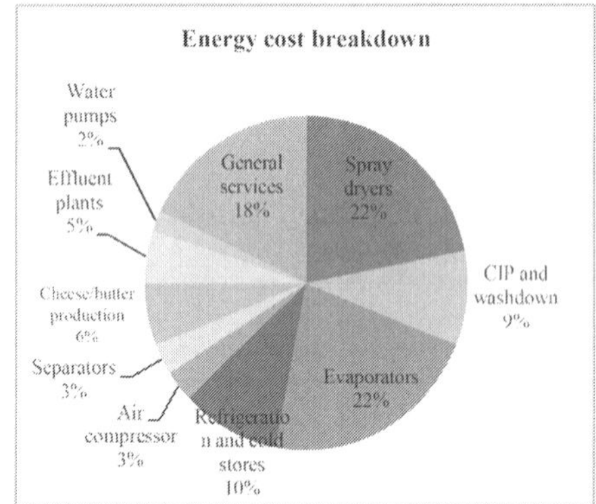

Fig. 3: Water Consumption and Energy Utilisation in a Typical Dairy Plant (Rad and Lewis, 2014)

Dairy industry is also being held responsible for generation of effluents having high organic load. The effluents are mostly composed of milk solids, which are mainly derived from various by-products like whey which is produced during manufacturing of cheese and paneer. Also, if the milk gets spoiled (may be due to high acidity), it is being thrown out of the plant and it forms a part of the effluent. The effluent also comprises of discarded rinse waters, which consist of acid and alkalis, used during CIP. Biological oxygen demand (BOD) and chemical oxygen demand (COD) has been directly related to waste generation. It has been observed that, waste water from a typical dairy industry has a 7.2 pH, alkalinity (600 mg/l as $CaCO_3$), total dissolved solids of about 1060 mg/l, suspended solids (760 mg/l), BOD (1240 mg/l), COD (84 mg/l), total nitrogen (84 mg/l), phosphorus (11.7 mg/l), oil and grease (290 mg/l) and chloride (105 mg/l) (Singh *et al.*, 2014). These dairy effluents being rich in organic and suspended solids, gets degraded with passage of time and releases gases, odour, adds colour and turbidity and paves the way for eutrophication (Singh *et al.*, 2014). As a result, if the effluents are discharged from the plant, without any treatment, it would lead to enormous environmental pollution.

Packaging materials of dairy products is another segment, which plays a part in greenhouse gas emissions. According to India Brand Equity Foundation (IBEF), 2009, about 49% of the packaging materials are made of plastics, followed by paper and paper board (12%), metal (7%), glass (8%) and others (24%). Plastic is thus significantly used in dairy packaging industry because of the rising demand of flexible packaging (lighter in weight, requires less storage space) as compared to rigid packaging. Most of the plastic materials are composed of synthetic polymers like polyethylene (PE), polypropylene (PP), polyvinylchloride (PVC), polyester (PS) and polytetrafluroethylene (PTFE), which are obtained from petrochemicals. Synthetic polymers are mostly preferred because of their strength, flexibility and chemical inertness. However, as these synthetic polymers are made from non- renewable resources like oil and natural gas, therefore, it leads to the depletion of these natural reserves. Also, due to the resistance of plastics (derived from synthetic polymers) to chemical and biological degradation, it may take almost 500 years to get degraded. Also, if these are incinerated, they lead to GHG emissions and thus lead to global warming. High Density Polyethylene (HDPE), Low density Polyethylene (LDPE), polypropylene (PP), polyvinylchloride (PVC), polyethylene terephthalate (PET), corrugated cardboard and wood has a carbon footprint of 1.65, 2.27, 2.02, 2.66, 3.77, 0.957 and 0.065 kg CO_2 equivalent/ kg, respectively (Mengual *et al.*, 2014). Recyclability of packaging materials also plays a role in carbon footprinting.

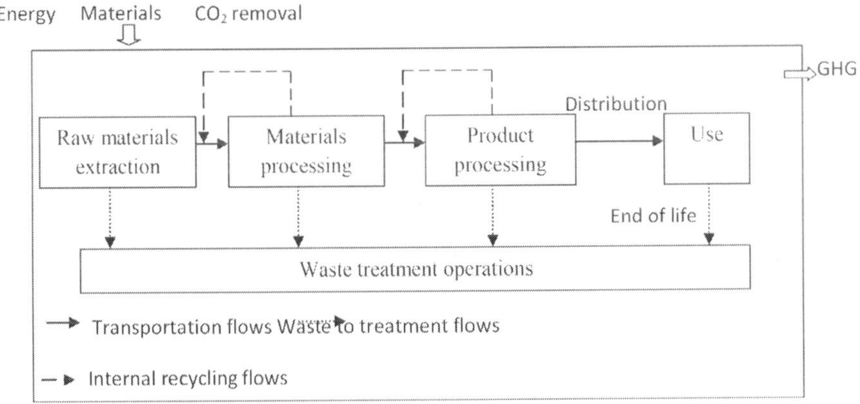

Fig. 4: LCA of a packaging material (cradle to cradle approach) (Mengual *et al.*, 2014)

14.5 LCA of Milk Transportation

Milk and milk products are transported over long distances, for the delivery of the product to retail points and consumers. The distribution of milk and milk products is quite challenging, considering the perishability and sanitary requirement of dairy products. Special packaging are used for transportation of milk products, like, milk crates made of HDPE are used for distribution of toned milk. Milk products are mostly transported by refrigerated or insulated trucks, however, other modes like ships, trains and planes can also be employed. The fuel utilized in transportation, forms the basis of greenhouse gas emissions. Transportation by air mode emits the highest amount of GHG (40 times more), followed by trucks (10 times more), railways and the least amount of GHG is emitted by ships. Thus, faster is the mode of transport, higher is the quantity of GHG produced per ton-km. Apart from the mode of transportation, time taken to deliver the product to consumer and conditions of transportation (i.e. temperature, for example, frozen dairy products are transported at freezing temperatures (-18°C) whereas transportation of milk powder does not require any refrigeration), also has a significant impact on GHG emissions. The impact of transportation on greenhouse gas emissions is generally below 5%, but it may be above 50% depending upon the products and processes (Boye and Arcand, 2012). The greenhouse gas emissions of the various steps involved in the production and processing of fluid milk and yoghurt has been specified in Table 3.

Table 3: Greenhouse gas (GHG) emissions (% of total) of dairy sector

Step	Fluid milk	Yoghurt
Milk Production at dairy farm	86.9	72.2
Transportation	1.0	0.9
Processing	6.5	16.8
Packaging	5.5	10.1
Total	**100**	**100**
Processing (Detailed)		
Electricity	28.5	41.0
Fossil fuels	66.6	57.3
Water and waste water	1.6	0.7
Cleaners	1.5	0.3
Refrigerants	1.9	0.7
Total	**100**	**100**

(Verge *et al.*, 2013)

14.6 Green Technology in Milk Production

Greenhouse gases can be reduced by the genetic improvement of the breeds. Genetic improvement can be achieved via economic breeding index (EBI) or by crossbreeding of indigenous cows with exotic breeds having superior milk yield. Genetic advancement increased the milk productivity by 5 to 8 times, shortened the calving intervals and also led to improvement in the health of cows (Brien *et al.*, 2016). Higher EBI, therefore, helps to reduce methane emissions from enteric fermentation, by increasing the yield of milk solids, without any deterioration in animal health. The Indian government has already put a lot of effort in this aspect. The frequency of artificial insemination (AI) increased by 124.1% within a tenure of ten years (2005-15). This paved the way for improved milk productivity in crossbreed and indigenous cattle at the rate of 1.1% and 2.7%, respectively during the same period (2005-15) (DADF, 2016). Also, government has increased the number of veterinary dispensaries and clinics throughout the country, for decreasing the disease outbreaks in cattle. Apart from genetic improvement, the diet or feed intake of the cattle also has a huge impact on the carbon footprint of milk. The greenhouse gas emissions can be reduced by raising the amount of pasture intake or grazed grass intake. In addition, higher amount of grazed grass in the feed decreases the storage requirements of manure and thereby lowers the CH4 emissions. This is because amount of manure excreted inside the cattle shed gets lowered and also lower amount of forage has to be harvested mechanically. As a result, extending the grazing season significantly lowers the GHG emissions and thus the carbon footprint of milk.

The nitrous oxide emission can be lowered by reducing the amount of fertilizer used and by increasing the addition of manure in the soil. The plant biomass (stems) remaining after the crops are harvested is a very rich source of CO_2 and starch. The high amount of carbon dioxide which is present in the plant bio-mass can be trapped or captured and this CO_2 can be used for preparation of various carbonated food and dairy beverages. The starch present in plant bio mass can be utilized for the production of bio-ethanol, which can be used as a fuel in automobiles (Classen *et al.*, 1999). In addition, the manure is a rich source of methane, which can utilized for production of bio-gas. Bio-gas is mainly composed of CH_4 (50-75% by volume), CO_2 (25-40% by volume), along with negligible quantity of hydrogen sulfide, nitrogen and oxygen and can be manufactured by anaerobic digestion of manure (Khanal *et al.*, 2019). This biogas can be employed as a fuel for vehicles, boilers (for producing hot water or steam) and it can also serve as a cooking medium (Suárez *et al.*, 2019).

Organic milk and milk products which has a low carbon footprint and wide consumer acceptability can be produced only in an organic dairy farm (Oruganti, 2011). Organic dairy farming is a concept of rearing animals on organic feed (i.e. feed which is produced without utilizing any fertilizers or pesticides). Even the land which is used for organic crop cultivation should be free from any prohibited materials for minimum three years, before the first harvesting of organic crops. The cattle's aged above six months must have access to the pastures. The animal feed should be free from genetically modified organisms (GMOs), solvent extracts, urea and slaughter by-products. Only organic milk should be fed to the calves. In addition, there should be limited usage of antibiotics (only to be used if animals are ill) and hormones (like oxytocin which is used for increasing milk productivity). Even when the animals are on antibiotics, milking of those diseased animals should be restricted and the withdrawal period should be higher in comparison to conventional dairy farming.

14.7 Green Technology in Milk Processing

Milk is processed for ensuring safety and for enhancing its keeping quality and to manufacture different value added dairy products. Thermal treatments like pasteurization and sterilization are the most conventional modes of inactivation of micro-organisms in milk and milk products. But these processes utilizes high temperature for microbial inactivation, which eventually lead to denaturation of heat sensitive components, thereby lowering the nutritive value and affecting the organoleptic properties of dairy products. In addition, thermal treatments consume a considerable amount of energy.As a result, non-thermal novel processes, are gaining importance because of minimal nutrient loss, lower energy consumption and also due to negligible changes in the organoleptic property of the product. Novel green technologies include high hydrostatic pressure (HHP), pulsed electric fields (PEF), ohmic heating (OH), microwave heating and treatment with ultrasonics and ultraviolet light (Barba *et al.*, 2019). However, most of these novel technologies are not widely used in dairy industries still now.

Drying and evaporation are the most energy intensive processes in dairy industries, which are utilized to manufacture condensed milk, skim milk powder, whole milk powder, whey powder, lactose and many others. Energy consumption can be minimized by utilizing membrane filtration technology, which includes, reverse osmosis, ultrafiltration, nanofiltration and micro-filtration. Reverse osmosis is mostly employed to concentrate the milk by removing the water and this process can be operated at ambient temperature. Ultrafiltration is utilized to separate protein from the milk or whey and can be used to manufacture whey protein concentrate. Microfiltration is mostly used for bacterial reduction of milk and whey; therefore it can serve the function of a milk clarifier or bactofugation unit. In addition to membrane technology, energy conservation in a milk evaporator section can be achieved by using multiple effect evaporators (as a replacement of single effect evaporator) along with a thermocompressor or a mechanical vapour recompression (MVR) unit. In dryer section, energy wastage can be minimized by using three stage dryer (as a replacement of single stage dryer) i.e. spray dryer integrated with fluidized bed dryer (because of better heat recovery). Heat can also be recovered from the warm exhaust gases and can be re-circulated in the dryer. In addition, fine milk solids have a tendency to escape from the dryer, by mixing with the exhaust gases, which leads to product losses. By utilizing cyclone separators or bag filters, fine milk solids can be recovered and the product wastage can be minimized.

Enzymes also play a crucial role in lowering the carbon footprint of dairy industry. Food enzymes, being biological catalysts, accelerate the reaction rate by lowering the activation energy required. Food enzymes are specific and non-toxic in nature and can be easily inactivated. Enzymes such as lipase, protease, esterase, lactase and transglutaminase are used in dairy industry for flavour enhancement (in cheese and ghee), accelerated ripening (in cheese), protein cross linking (provides better body and texture in yoghurt) and lactose hydrolysed dairy products (suitable for lactose intolerant people) (Abada, 2019). However, the major usage of enzyme in dairy industry is for manufacturing of cheese, which is a rennet coagulated product. Flavour development in cheese mostly occurs during ripening by proteolysis, lipolysis and glycolysis. Ripening process is an energy intensive process as cheese is kept at a low temperature for a long period of time, for example, optimum flavour development in cheddar cheese occurs after ripening it for six to nine months. Lipase, esterase and protease enzyme has been added to accelerate the ripening, as these enzymes enhance the flavour development by lipolytic or proteolytic pathway.

To reduce the wastage of milk solids in dairy industry, it is imperative to utilize the by-products and convert it into value added dairy products. The major by-products of dairy industry are skim milk (produced during manufacturing of cream), whey (obtained during preparation of chhana, paneer and cheese), butter milk (derived during butter manufacture) and ghee residue (produced during ghee manufacture).Casein, caseinate and co-precipitates can be derived by means of acid (hydrochloric/lactic acid) or rennet precipitation of skim milk. Casein, caseinates and co-precipitates can be utilized to increase the protein content in various dairy and food products. In addition, casein can be used to manufacture adhesives and plastics. Whey can be used to prepare whey beverages (for example, Amul Tru commercially available in Indian market), whey powder, whey protein concentrate (WPC), whey protein isolate (WPI), ethanol, bio-gas, lactose powder and various other fermented whey products (single cell protein, baker's yeast etc.). WPC has got huge commercial importance as it is widely used by athletes and body builders. Whey beverage serves the dual purpose of thirst quenching and rehydration solution (as whey is rich in electrolytes). Whey powder is primarily used to make up the requirement of SNF (solids-not- fat) in various dairy products and it is also utilized to enhance the protein content in various food products. Lactose is commercially utilized in pharmaceutical application (as a filler and bulking agent), bakery products (to impart better flavour and colour) and confectionary products (improves body and texture of the product). Buttermilk can be consumed as a thirst quenching beverage in summer season and it has also got various health promoting attributes (like it improves digestion). Buttermilk powder can also be manufactured which helps to make up the total solid (TS) content of dairy product. Buttermilk powder (BMP) has been added in yoghurt to prevent syneresis as BMP is very rich in phospholipids and it has got water binding ability. Ghee residue is utilized to manufacture various confectionary and bakery products.However, the milk solids which get lost because of washing of process lines and cans or due to leakage in process lines are generally treated in effluent treatment plant (ETP) before discharging it from the plant. ETP also treats all the discharged rinse water,which consists of chemicals (acids and alkalis), thus it reduces the organic load and suspended solids in the discharge water and eventually the BOD and COD, thereby decreasing the environmental pollution (Singh *et al.*, 2014). Also, as the dairy industries utilize a large amount of water; many dairy industries are harvesting rain water, such as to reduce the usage of fresh water.

As most of the dairy products are perishable in nature, therefore chemical preservatives are added to enhance the product's shelf life. However, these chemical preservatives may have serious health implications. Bio-preservatives can serve as a replacement of these chemical preservatives. Bacteriocins, particularly, nisin is the most widely used bio-preservative in dairy and food industry. Nisin (commercially available as Nisaplin) is recognized as a GRAS (Generally recognized as safe). Nisin has been added in pasteurized milk and processed cheese, for restricting the growth of *Clostridium tyrobutyricum*, which remains active even after severe heat treatments (85-105°C) (Sen and Ray, 2020). Nisin also plays a preservative role in canned evaporated milk and in various dairy desserts, where sterilization is not done as it would lead to poor body and texture and appearance of the product.Packagingmaterials has also been observed as a serious environmental pollutant. As plastic packaging materials takes a very long time to degrade, therefore the concept of bio-degradable packaging emerged which degrades comparatively in a short period of time. Bio- degradable packaging materials are mostly made from renewable and more sustainable sources like biomass such as starch, cellulose, chitin, casein and whey protein. Bio-degradable packaging materials include polylactic acid (PLA), bio-polyethylene (bioPE), polyhydroxyalkanoates (PHA), poly (carpolactone) (PCL) etc (Robertson, 2016). However, due to superior properties of synthetic polymers like PP, PE etc. and due to lower cost, bio-degradable packaging materials are still now not widely used in dairy industries. Apart from bio-degradable packaging, many dairy industries are utilising digital mode or digital library to store all their records or data, thus minimizing the usage of paper and reducing the carbon footprint.

14.8 Conclusion

In dairy sector, milk production (enteric fermentation)in dairy farm is the major source of greenhouse gas emissions followed by milk processing, packaging and transportation.GHG emissions from milk production can be reduced by lowering the usage of fertilizers, increasing the grazing season and by utilizing the plant biomass (rich in CO_2) and manure (rich in CH_4) for preparation of bio-ethanol and bio-gas.The carbon footprint of the milk processing sector can be minimized by using novel non- thermal technologies (HPP, PEF etc.), membrane filtration, enzymes, bio-degradable packaging materials, bio-preservatives and by means of reducing the wastage by utilising the by-products. Also, wastes should be treated in an effluent treatment plant (ETP) prior to discharge. However, most of the green technologies cannot be applied on industrial scale. Therefore, considerable amount of research is needed to develop various green or eco friendly technologies so that they can be utilized on a commercial scale, while keeping in mind the economical point, environmental impact and consumer's health.

14.9 References

Abada EA. 2019. Application of microbial enzymes in the dairy industry:Enzymes in Food Biotechnology, pp. 61-72, Academic Press.

Arcand Y, Maxime D, Zareifard R. 2012. Life cycle assessment of processed food : Green Technologies in Food Production and Processing, pp. 115-148, Springer.

Barba FJ, Rosello Soto E, Marszałek K., Kovacevic DB, Jambrak AR, Lorenzo JMand Putnik P 2019. Green food processing- Concepts, strategies and tools : Green Food Processing Techniques, pp. 1-21, Academic Press.

Boye JI and Arcand Y. 2013. Current trends in green technologies in food production and processing. Food Engineering Reviews, 5(1), 1-17.

Claassen PAM, Van Lier JB, Contreras AL, Van Niel EWJ, Sijtsma L, Stams AJMand Weusthuis RA. Utilization of biomass for the supply of energy carriers. Applied microbiology and biotechnology, 52(6), 741-755, 1999.

Department of animal husbandry and dairying, https://dahd.nic.in/, 2016.

Federation of Indian Chambers of Commerce and Industryhttps://www.google.com/search?rlz =1C1RLNS_enIN910IN910&sxsrf=ALeK k03Mh7FNxHFK2zLr2O PWiTGO 6iuw %3A 1603045985293&ei=YYqMX6W4EYn5 9QPn0aGYCg &q =Indain+ dairy+ association+45th+dairy+industry+conference+nddb+ coop+pdf&oq =Indain+ dairy+association+45th+dairy+industry+conference +nddb+coo p+pdf&gs_lcp=CgZwc3 ktYWIQAzIHCCEQChCgATIHCCEQChCgAToECAAQR1

DEFFi2OmCXPmgAcAF4AIAB0wuIAdQZkgEHNS0xLjAuMpgBAKABAaoBB2d3 cy13aXrIAQjAAQE&sclient=psyab&ved=0ahUKEwjl7IvU477sAhWJfH0KHedoCK MQ4dUDCA0&uact=5#, 2020.

Garg MR, Phondba B, Sherasia P. 2018. Carbon footprint of Milk in India: Past Trends and Future Prospects. Indian Dairyman, 70(1), 72-78.

India Brand Equity foundation, 2009. https://www.ibef.org/download/Flexible_Packaging 060112.pdf.

Indian Dairy Association, 45th Dairy Industry conference, https://www.nddb.coop/about/ speech/dic, 2017.

International Dairy Federation, https://www.fil-idf.org/wp- content/uploads/2016/09/Bulletin 479-2015_A-common-carbon-footprint-approach- for-the-dairy-sector.CAT.pdf, 2015.

Khanal SK, Nindhia TGT, Nitayavardhana S. 2019. Biogas from wastes, processes and applications: Sustainable Resource Recovery and Zero Waste Approaches, pp. 165- Elsevier.

National Accounts Statistics, http://mospi.gov.in/publication/national - accounts- statistics, 2019.

National Dairy Development Board, https://www.nddb.coop/, 2020.

O'Brien D, Geoghegan A, McNamara K and Shalloo L. 2016. How can grass-based dairy farmers reduce the carbon footprint of milk?. Animal Production Science, 56(3), 495- 500.

Oruganti M. 2011. Organic Dairy Farming-A New Trend in Dairy Sector. Veterinary World, 4(3), 128.

Rad SJ, Lewis MJ. 2014. Water utilization, energy utilization and waste water management in the dairy industry: A review. International Journal of Dairy Technology, 67(1), 1- 20.

Ramos-Suarez JL, Ritter A, González JM and Pérez AC. 2019. Biogas from animal manure: A sustainable energy opportunity in the Canary Islands. Renewable and Sustainable Energy Reviews, 104, 137-150.

Robertson GL. 2016. Food packaging: principles and practice, pp.200-210, CRC press.

Sanyé-Mengual E, Lozano RG, Oliver-Solà J, Gasol CM and Rieradevall J. 2014. Eco-design and product carbon footprint use in the packaging sector: Assessment of Carbon Footprint in Different Industrial Sectors, Volume 1, pp. 221-245, Springer.

Sen C, Ray PR. 2019. Biopreservation of Dairy Products using Bacteriocins. Indian Food Industry Magazine,1(4),51-60.

Singh NB, Singh R, Imam MM. 2014. Waste water management in dairy industry: pollution abatement and preventive attitudes. International Journal of Science, environment and technology, 3(2), 672-683.

Vergé XP, Maxime D, Dyer JA, Desjardins RL, Arcand Y and Vanderzaag A. 2013. Carbon footprint of Canadian dairy products: Calculations and issues. Journal of Dairy Science, 96(9), 6091-6104.

Colour Plates

Chapter 1: Green Technology in Bakery Industry

Fig.1: Bakery Machinery

Fig. 2: Biogas plant

Fig. 3: Solar panel

Chapter 7: Valorization of By Products in Agro Commodity Processing Sectors

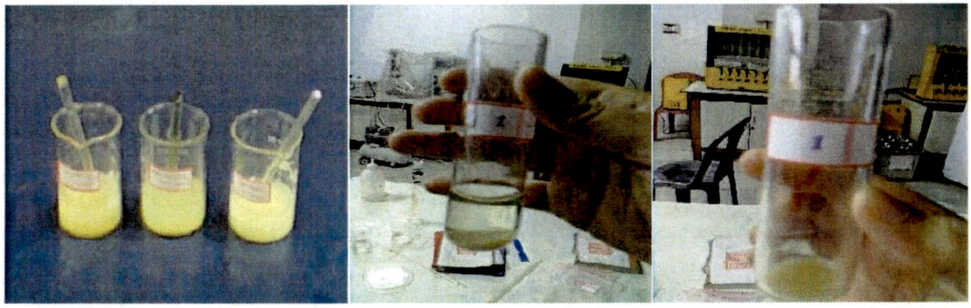

Swelling power = Weight of sediment paste x 100/Weight of sample on dry basis x (100 -% solubility) (1)

Chapter 11: Cooked Food Waste Management An Eco-friendly Approach

Fig. 1: Bokashi method (*Source*: Ambong *et al.*, 2018)

Chapter 12: Green Packaging and Utilization of Fruit Fibres: A Review

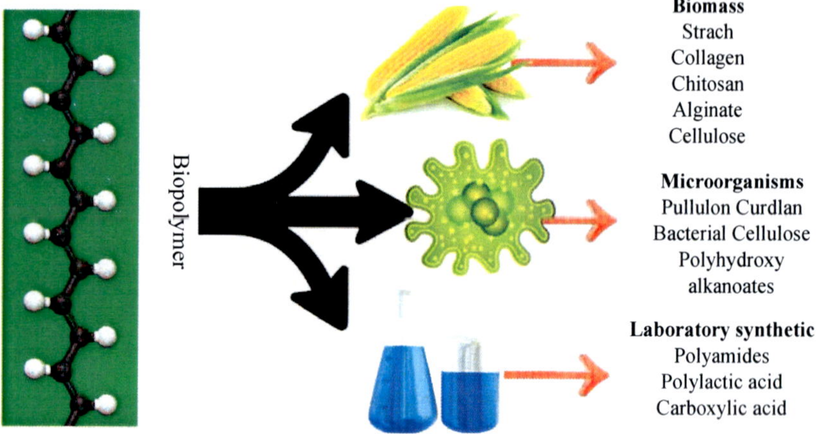

Fig. 2: Types of Biopolymers on the basis of sources (Qasim *et al.*, 2020)

Fig. 4: Pineapple Leaf Fibres

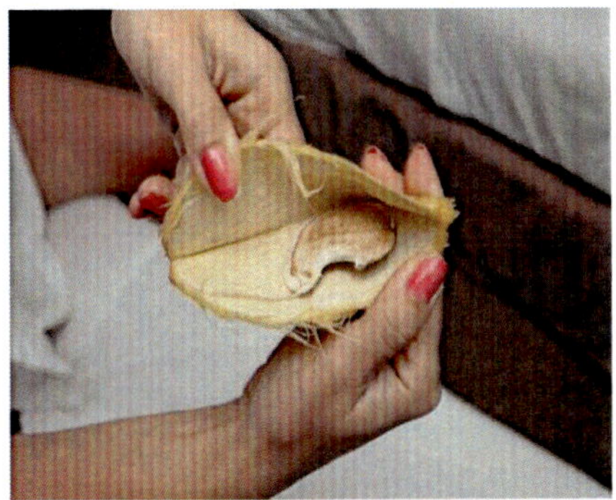
Fig. 5: Mango Kernel

Fig. 6: Coconut Fibres

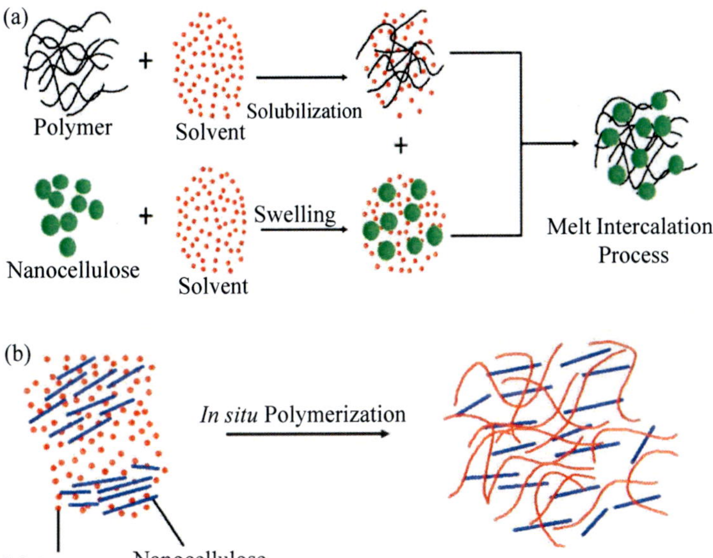

Fig. 7: Manufacturing of Nanocomposites (Qasim *et al.*, 2020)

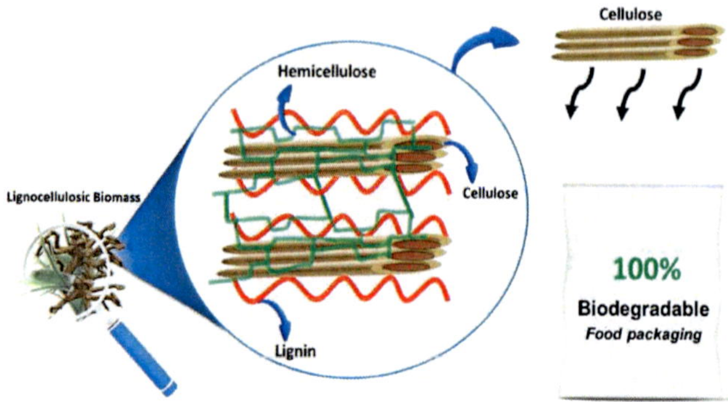

Fig. 9: Cellulose in Green Packaging (Qasim *et al.*, 2020)

Chapter 13: Processing of Banana Flower to Value Added Products: A Review

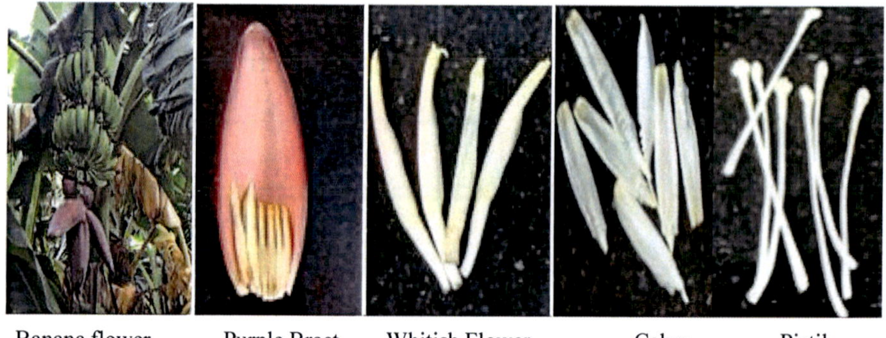

Fig. 1: Different parts of the banana flower

Fig. 2: Image of marketed dried banana flower

Fig. 3: Image of marketed dried banana flower